SQL++ For SQL Users:
A Tutorial

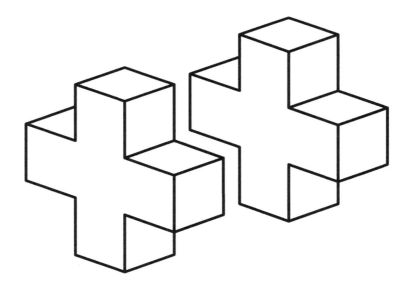

Don Chamberlin

September 2018

SQL++ For SQL Users: A Tutorial by Don Chamberlin

Copyright © 2018 by Couchbase, Inc.

For information about this title contact:
Couchbase, Inc. | 1-650-417-7500 | Couchbase.com

978-0-692-18450-9

Printed in the United States of America

Cover and Interior Design: All Around Creative, Inc.

Couchbase

START A REVOLUTION

couchbase.com

Foreword

by Michael J. Carey, Ph.D., UC Irvine

It is a pleasure to introduce this new book by Don Chamberlin, as I have known Don for many years and his writings are always exceptional. I first came to know Don personally during several sabbatical visits from the University of Wisconsin to the database research group at IBM Almaden Research Center. I later had the pleasure of working with Don as an IBM colleague on object-relational extensions to SQL (SQL:1999 and SQL:2003) and to DB2, and later still on XQuery, the W3C query language standard for XML data, when I was doing information integration work at BEA Systems and he was representing IBM Research center in San Jose, in the brave new world of XML and XML queries. My experience with Don is that, in addition to being way too modest for someone who is the IBM Fellow who co-invented SQL and the world's leading expert on query language design, he asks tough questions that get to the heart of problems, illustrates things in the clearest possible terms using insightful examples, and has a way of explaining things to make them readily accessible to readers at all levels. In my opinion, Don's book "A Complete Guide to DB2 Universal Database" from the 1990s was, and still is, one of the best books ever written for someone who wants to learn about SQL and its realization in IBM's DB2 system for Unix- and Windows-based platforms.

In my role as a part-time consultant to Couchbase, Inc., I was excited when another chance to work with Don came about: Don's interest in query languages somehow landed him at an early Couchbase seminar on the first version of N1QL, an SQL-inspired language for JSON data, and he became intrigued. Meanwhile, my work as an architect for Apache AsterixDB and Couchbase Analytics put me in a position to be intrigued by SQL++, a SQL for JSON query language initially designed by Yannis Papakonstantinou and students at University of California, San Diego. I was beginning to tire of trying to convince our AsterixDB users that its initial query language (AQL, inspired by XQuery) was what they should really want for semi-structured data, Yannis had a thoughtfully designed alternative (and Couchbase was working with UCSD on N1QL), and Don was freshly retired from IBM but not from thinking about query language design. We joined forces and, together with some technical leads at Couchbase and elsewhere, SQL++ has been refined into a potential standard for querying JSON and JSON-like data.

Enough background! The world of JSON and "NoSQL" brings new challenges for querying: schemas are optional and often absent, and objects are nested and often heterogenous in structure. SQL++ addresses those challenges, and this book provides a gentle yet thorough introduction to SQL++ for analysts and developers who have a working knowledge of SQL.

In particular, this book does a wonderful job of highlighting the key data model and language differences that JSON brings and the resulting potential pitfalls—and avoidance patterns—that someone new to SQL++ should watch out for. An early draft of this book has been "beta tested" on over 500 students in database classes at the University of California, Irvine and at the University of Washington, so I know first hand that it meets its objectives. Read on and I predict that you'll soon see why Don is "the man" when it comes to books of this sort. Enjoy!

Table of Contents

Preface

Preface

SQL++ is a query language for semi-structured data based on the JSON data format. It was developed by Professor Yannis Papakonstantinou and others at University of California, San Diego (UCSD) (see details at https://arxiv.org/abs/1405.3631). As described in UCSD publications, SQL++ is a parameterized language, with a syntax similar to that of the well-known SQL query language, and with parameters that control certain aspects of its semantics such as how to deal with null or missing values and other aspects pertaining to SQL extensions.

Couchbase is a database management system with a distributed architecture. It is designed for web-scale applications where performance, scalability, and availability are critical. Couchbase Server stores JSON documents and supports a query language called N1QL (pronounced "nickel"), designed by Gerald Sangudi and others at Couchbase. The name N1QL suggests "not first normal form," a reference to the fact that JSON documents, unlike relational data, may contain nested data structures.

Couchbase is tuned for high performance and low latency on business-critical data. Data stored in Couchbase can be accessed and updated through various APIs and through a Query Service using the N1QL language. In addition, Couchbase supports a separate Analytics Service for long-running or ad hoc queries that would not be appropriate to run directly on operational data. The latest operational data is automatically and continuously replicated on the Analytics Service, where it can be queried without impacting the performance of real-time business operations. The Analytics Service is based on technology co-developed by the Asterix research project at University of California, Irvine and University of California, Riverside, directed by Professor Michael Carey (see details at http://asterix.ics.uci.edu/).

The Query Service and the Analytics Service use slightly different versions of the N1QL language. Where the distinction is important, we refer to these versions as N1QL for Query and N1QL for Analytics, respectively. N1QL for Analytics is the Couchbase implementation of SQL++, with some additional features for compatibility with earlier versions of N1QL. Since this tutorial is about SQL++, all the example queries in this document are written in N1QL for Analytics and tested on the Analytics Service. Appendix C contains instructions for how you can load the example data into Couchbase and run the example queries yourself.

In the remainder of this tutorial, the term N1QL refers to N1QL for Analytics unless specified otherwise.

The syntax diagrams and descriptions in this document are designed to serve as an introductory tutorial rather than as a complete technical reference. In some cases, complex optional clauses or alternative keywords have been omitted from the syntax. For a complete technical reference, see the official Couchbase documentation.

Introduction

Introduction

If you know something about SQL and want to leverage that knowledge to gain a working knowledge of SQL++, this tutorial is for you. I'll assume that you are familiar with the basic SELECT-FROM-WHERE structure of an SQL query. I'll assume that you know a little bit about joins and grouping, as defined in the SQL Standard and supported in all SQL implementations. I won't expect any knowledge of advanced or complex SQL features.

The good news is that many SQL++ queries will look quite familiar to you. For example, consider this query, which might be executed at an insurance company:

```
SELECT policy_number, monthly_premium, claims
FROM policies
WHERE monthly_premium >= 1000
ORDER BY monthly_premium DESC
LIMIT 2;
```

This query should look perfectly familiar to you. It might return the following result:

```
[ { "policy_number": "18537",
    "monthly_premium": 1260,
    "claims": 2
  },
  { "policy_number": "21640",
    "monthly_premium": 1025
  }
]
```

This result should be easy for you to understand, but you may have some questions about it. First of all, it doesn't look like a table because (among other things) one policy has a value for "claims" and the other policy does not. You'll soon see that, in SQL++, data (including query results) are not required to have rows and columns. Not being limited to a conventional concept of tables is the first and possibly the most difficult change that you will have to face.

As an SQL user, you are used to working with tables, which have a very regular structure. In a table, every row looks pretty much the same—they all have the same columns, and each column has the same type in every row. Even when data is missing—for example, when a customer has no middle name—the "middle name" column in the customer table is still there. That's why null values were invented.

The regular structure of a table has many advantages for a database system. It means that the names and types of columns only need to be recorded once, not repeated on every row, because all the rows are the same. It also means that the type of every value stored in the table is known in advance. If the `price` column of a table is defined to be of type INTEGER, then the `price` value in every row of the table will be exactly one integer (or null); never a string or an array of integers or a nested table. This makes it easy for an optimizer to create an efficient plan for searching the table. Tables are also good for "normalization," a database design discipline that ensures that each fact is stored exactly once, eliminating redundancy and inconsistency. These advantages are some of the reasons why tables have been the predominant format for business data for the last 40 years.

If you are storing data for hundreds of thousands of bank accounts, and all the bank accounts look pretty much the same, tables are the data structure for you. Lots of data fits nicely into this format. Tables are not about to go away.

But suppose that you need to store data about car insurance policies. In your business, not all car insurance policies look the same. They all have some parts in common, such as policy number and billing address. But you do business in several states, and each state has its own rules about how car insurance coverage needs to be organized. Some are "no fault" states and some are not. Some policies include collision coverage and some do not. Some cars have only one driver, and some have a list of several drivers, with data on each driver (which may also vary by state). Some policies (your favorite ones) have no accidents on record, and some have lists of accidents. For each accident, you may have no witnesses or you may have a list of witnesses, and the witness reports might include text strings, images, and PDF files.

Relational databases are capable of storing your insurance policy data. To do so, they will split up the data across many tables—a policy table, a driver table, several kinds of coverage tables, an accident table, a witness table, and so on. Each of these tables will have a schema that defines what kinds of data values it can hold. There may be "referential integrity" rules that govern how the rows of one table are related to the rows of other tables. All these rules and schemas need to be checked and enforced by the database system. When a query depends on several facts, such as "Find policies in which a non-principal driver has had more than one accident," information from multiple tables must be joined together to find the answer. The process of enforcing rules and schemas, and joining data from multiple tables, costs machine cycles and may impact the performance of your application.

The appeal of schemaless, or "NoSQL," systems is that they relax many of the constraints that are enforced by relational systems. In a NoSQL system, data doesn't have to look like tables. All the information about one insurance policy can be stored in one place, even though different policies have different kinds of data, and even if the policies contain some "repeating groups." Queries against stored data may be simpler to write and faster to execute because they don't need so many explicit joins.

Ironically, SQL++, which is based on SQL, is often considered to be a member of the "NoSQL" class of languages because it does not require data to be constrained to a tabular format. Rather than the relational model of data, SQL++ uses a more flexible data model based on JSON (JavaScript Object Notation), a very simple notation that is widely used for exchanging data between web applications.

At this point, you may be thinking, "How can SQL++, which is based on SQL, get along without tables, when SQL was specifically designed to query tabular data?" I have two answers for you: (1) Surprisingly well; and (2) Mainly by relaxing some of the constraints and limitations of original SQL. After reading the rest of this tutorial, I'm hoping you will agree. We'll explore this topic using a series of examples written in the Couchbase implementation of SQL++, called N1QL for Analytics, abbreviated here simply as N1QL.

JSON

JSON

JSON (JavaScript Object Notation) is a simple format for storing and exchanging structured data. It's easy for humans to read and write, and it's easy for machines to parse and generate. It is based on name-value pairs, which makes data self-describing. It can be represented as a character string, which makes it easy to exchange between a browser and a server. It can be easily converted to and from the native data structures of the JavaScript programming language. For all these reasons, JSON is a popular format for e-commerce and other web-based applications.

All data operated on by N1QL, and all query results generated by N1QL, are JSON values. A JSON value may be either a primitive value or a structured value. A primitive value is any of the following:

- A string, which looks like `"Hello"`
- A number, which looks like `42` or `-3.14159`
- `true` or `false`
- `null`

A structured value is any of the following:

- An object, which is a list of name-value pairs enclosed in curly braces and separated by commas. The name-value pairs are called *fields*. Each name is a string, and each value may be any JSON value. Here's an example of an object that contains two fields:

  ```
  { "partno": 461,
    "description": "Wrench"
  }
  ```

  JSON does not define an ordering among the fields of an object.

- An array, which is an ordered list of items enclosed in square brackets and separated by commas. Each item may be any JSON value. Here's an example of an array:

  ```
  [ 1, 2.5, "Hello", true, null ]
  ```

Since the items in an array, and the values in the fields of an object, can be any JSON values, arrays and objects can be nested inside each other without any restrictions. The items in an array, or the field-values in an object, need not all have the same type.

JSON itself does not have a concept of a collection of objects that all have the same fields like rows of a table. A database designer who is used to working with tables might choose to create such a collection, but the uniformity of the objects is not required or enforced by the database system.

For some applications, a database designer might choose to take advantage of JSON's flexibility to store heterogeneous data. Consider an object that represents a book and has an author field. Some books have one author, other books have many authors, and a few books have no authors listed. Since JSON is a schemaless data format, it's possible for the author field to sometimes contain a string, sometimes contain an array of strings, and sometimes be missing entirely. This is valid JSON but it may not be a good database design. This design makes it difficult to write a query that returns the authors of a given book, because it's not known whether the query needs to unpack an array or not. A better approach to situations like this is to use a consistent datatype for the author field, in this case an array that might contain zero, one, or many strings.

Arrays, Multisets, and Collections

As we've seen, a JSON array is an ordered collection of items, like [1, 2, 3]. But during processing of a query, it is sometimes necessary to deal with a collection of items that has no well-defined order. For example, a query block that has no ORDER BY clause returns an unordered collection of results. Recognizing that some collections have no order is important because it gives the query optimizer some flexibility in choosing an access plan. The query engine does not need to be careful to preserve the order of a collection if it has no order to begin with.

In this tutorial, we'll use the term *array* to denote an ordered collection, and the term *multiset* to denote an unordered collection. We'll use the generic term *collection* to denote something that may be either an array or a multiset. Many features of N1QL operate on collections. In a few cases, as we'll see, an operator requires its operand to be an array.

In our examples, we'll use JSON square brackets like [1, 2, 3] to denote any kind of collection.

An optimizing compiler will keep track of whether order is significant in each collection.

Of course, the complex data structures of JSON can be nested inside each other without any restrictions. The field-values of an object can be collections or other objects. The items in a collection can be objects or other collections. By combining these structures, you can build databases that are as complex as you like.

To facilitate interchange of data with other systems, N1QL always returns query results in JSON format. Multisets occur only in internal processing of queries. If the result of a query contains a multiset, it is converted into a JSON array with an arbitrary order.

Datasets and Dataverses

In N1QL for Analytics, the top-level things that can be directly queried in a database (like tables in SQL) are called *datasets*. A dataset has a name and a value. In N1QL, the value of a dataset is always a multiset of objects.

N1QL allows datasets to be organized into higher-level groups called *dataverses*. The name of each dataset must be unique within its dataverse. Dataverses may be helpful to separate the datasets used by different applications. For example, you might have a dataverse called inventory and another dataverse called development, and both of these dataverses might contain a dataset named products.

You may have noticed that a dataset that contains a multiset of homogeneous objects looks a lot like a table. Consider the following example:

```
[ { "partno": 1387,
    "description": "gear",
    "weight": 28.3 },
  { "partno": 1896,
    "description": "pulley",
    "weight": 19.7 },
  { "partno": 2114,
    "description": "belt",
    "weight": 13.5 }
]
```

This data might conform to a relational schema in which a table (unnamed, so far) has three columns named `partno`, `description`, and `weight`. That's why in many cases a N1QL query looks just like an SQL query. If the data shown above were contained in a dataset named `parts`, we could execute the following query:

```
SELECT description, weight
FROM parts
WHERE weight < 20;
```

and get the following result:

```
[ { "description": "pulley",
    "weight": 19.7
  },
  { "description": belt",
    "weight": 13.5
  }
]
```

The result of this query is a multiset of objects. Each "row" in the query result is represented by an object containing the names and values of the fields to be returned (in this case, `description` and `weight`). We'll see later how to write a query that returns a simple value like 47 that is not wrapped inside an object.

The final result of a query is always represented as a JSON array, whether its order is significant or not. If the query has an empty result, it is represented as an empty array, as shown in this example:

```
SELECT description, weight
FROM parts
WHERE weight > 100;
```

Result: [ ]

As you can see, tabular data can be stored in JSON format using datasets in place of tables and objects in place of rows. In this case, N1QL queries on tabular data look just like SQL queries.

But the most interesting parts of N1QL are about what happens when the stored data does not have a tabular format. That's what we will learn about in the following sections.

One difference that you'll probably notice between SQL and N1QL has to do with the ordering of fields in a query result. In SQL, this ordering is defined by the order in which the fields appear in the SELECT clause. For example, if you write SELECT description, weight FROM parts, you can expect each record in the query result to consist of a description followed by a weight. But since JSON does not define an ordering among the fields of an object, N1QL does not guarantee that the fields of the query result will appear in any particular order. In the above example, weight might appear before description. If you select many fields, some of which contain nested objects or arrays, your query result may be difficult to interpret. Control over the order of fields in a query result might be a reasonable thing to wish for in a future release.

In the examples in this tutorial, the order of fields in a query result always reflects their order in the corresponding SELECT clause. I've presented query results in this way to make them more readable. If you run the example queries on the Couchbase Analytics Service, the ordering of fields in your query results may differ from the order in which they appear in this book.

Expressions

Expressions

An expression is a piece of syntax that can be evaluated to return a value. N1QL supports nearly all of the kinds of expressions in SQL, and adds some new kinds as well.

As in SQL, any expression can be enclosed in parentheses to establish operator precedence; also as in SQL, subqueries must always be enclosed in parentheses.

In this section, we'll discuss each kind of N1QL expression, beginning with the simplest ones.

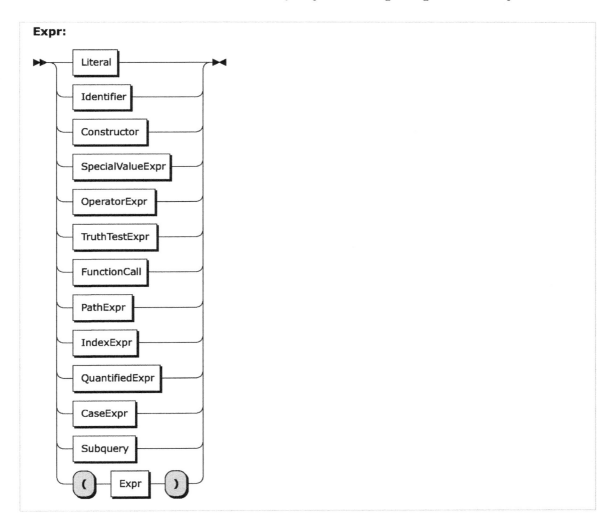

Literals

The simplest kind of expression is a literal that directly represents a value in JSON format. Here are some examples:

```
-42
"Hello"
true
false
null
```

Numeric literals may include a sign and an optional decimal point. They may also be written in exponential notation, like this:

```
5e2
-4.73E-2
```

String literals may be enclosed in either single quotes or double quotes. Inside a string literal, the delimiter character for that string must be "escaped" by a backward slash, as in these examples:

```
"I read \"War and Peace\" today."
'I don\'t believe everything I read.'
```

Identifiers and Variables

Like SQL, N1QL makes use of a language construct called an *identifier*. An identifier starts with an alphabetic character or the underscore character _ , and contains only case-sensitive alphabetic characters, numeric digits, or the special characters _ and $. It is also possible for an identifier to include other special characters, or to be the same as a reserved word, by enclosing the identifier in back-ticks (it's then called a *delimited identifier*). Identifiers are used in variable names and in certain other places in N1QL syntax, such as in path expressions, which we'll discuss soon. Here are some examples of identifiers:

```
X
customer_name
`SELECT`
`spaces in here`
`@&#`
```

A very simple kind of N1QL expression is a variable, which is simply an identifier. As in SQL, a variable can be bound to a value, which may be an input dataset, some intermediate result during processing of a query, or the final result of a query. We'll learn more about variables when we discuss queries.

Note that the N1QL rules for delimiting strings and identifiers are different from the SQL rules. In SQL, strings are always enclosed in single quotes, and double quotes are used for delimited identifiers.

Constructors

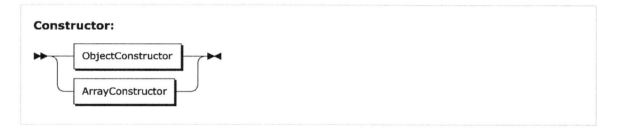

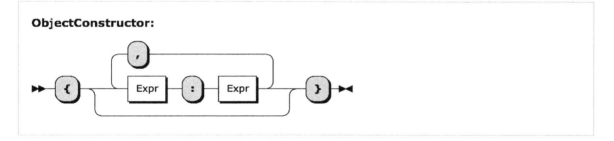

Structured JSON values can be represented by constructors, as in these examples:

```
An object: { "name": "Bill", "age": 42 }
An array: [ 1, 2, "Hello", null ]
```

In a constructed object, the names of the fields must be strings (either literal strings or computed strings), and an object may not contain any duplicate names. Of course, structured literals can be nested, as in this example:

```
[ {"name": "Bill",
    "address":
        {"street": "25 Main St.",
         "city": "Cincinnati, OH"
        }
    },
    {"name": "Mary",
     "address":
        {"street": "107 Market St.",
         "city": "St. Louis, MO"
        }
    }
]
```

The array items in an array constructor, and the field-names and field-values in an object constructor, may be represented by expressions. For example, suppose that the variables firstname, lastname, salary, and bonus are bound to appropriate values. Then structured values might be constructed by the following expressions:

An object:

```
{ "name": firstname || " " || lastname,
  "income": salary + bonus
}
```

An array:

```
["1984", lastname, salary + bonus, null]
```

Special Values

As you know, both SQL and JSON have a "null" value that is used to represent missing data. But JSON also has another way to represent missing data: it can be represented by data that is simply not present. For example, suppose that your database contains the following list of two customers:

```
[ {"first_name": "Mary",
   "middle_initial": "A.",
   "last_name": "Smith",
   "age": 65
  },
  {"first_name": "Bill",
   "last_name": "Wilson",
   "age": null
  }
]
```

You'll notice that Mary Smith has both a middle initial and an age, but Bill Wilson has neither. But the absence of an age is represented in the Bill Wilson object by a null value, whereas the absence of a middle initial in the same object is represented simply by the fact that there is no `middle_initial` field. Both of these representations are valid in JSON and in N1QL.

It's up to you, the user, to decide how you would like to use these two representations. One possible way is to use the null value to represent data that is applicable but unknown (for example, Bill Wilson has an age but we don't know what it is), and to use the absence of a field to represent data that is not applicable (for example, we know that Bill Wilson has no middle initial). N1QL doesn't force you to interpret the data in this way, but it is one reasonable way to use the facilities supported by the language and the data model, and it is compatible with the examples in this tutorial.

Since not all objects have the same fields, we need to define what happens when an expression references a field that is not present. For this purpose, N1QL defines a second special value, called `missing`, to represent the value of data that is simply "not there." If you reference a field that is not present in a given object, or an item that is outside the bounds of an array, you'll get the value `missing`.

N1QL provides NULL and MISSING keywords that you can use in expressions. For example, you can write `[0, 1, NULL, MISSING, 4]` to construct an array of length 5 in which the third item is null and the fourth item is missing (has no value).

Since the value of any expression can be `null` or `missing`, N1QL needs to define how these special values affect the various operators of the language. In general, if any operand of an expression is `missing`, the result is `missing`; otherwise if any operand is `null`, the result is `null`; otherwise the usual rules apply. However, the logical operators AND and OR have some special rules for handling `null` and `missing` (for these, see the section on Operator Expressions).

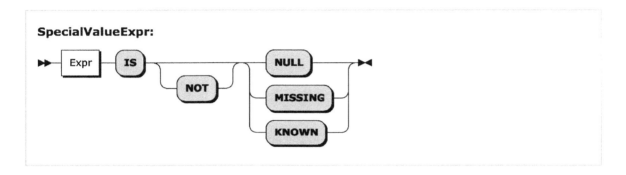

As you know, SQL has a special comparison expression IS NULL that returns `true` if its operand is `null`, otherwise `false` (and an IS NOT NULL expression that does the opposite). N1QL also supports the IS NULL and IS NOT NULL expressions, and also supports two additional new pairs of expressions:

- IS MISSING returns `true` if its operand is `missing`; otherwise it returns `false` (IS NOT MISSING does the opposite).

- IS KNOWN returns `true` if its operand has a value that is neither `null` nor `missing`; if the operand is `null` or `missing`, it returns `false`. (IS NOT KNOWN does the opposite.) You may also use the alternative keyword VALUED in place of KNOWN.

For the purpose of grouping, `null` and `missing` are considered to be separate values. If they occur in the grouping keys of a GROUP BY clause, a group will be created for each of these special values.

When encountered as an ordering key in an ORDER BY clause, `missing` is ordered before `null`, and `null` is ordered before any other value (if DESC is specified, the ordering is reversed).

The final result of a query is always returned in JSON format. Since the special values `null` and `missing` can pop up anywhere, we need to specify how they are serialized into JSON. Since `null` is a valid JSON value, that one is easy. But `missing` is not a valid value in JSON—it's just used as a placeholder for things that aren't really there. The rules for serializing `missing` values into JSON for query results are as follows:

- If the value of a field in an object is `missing`, that field is not serialized at all—it just doesn't appear in the output.

- If a `missing` value appears as an item in an array or multiset, it is converted to a `null` value in the serialized output.

If you were expecting a query to return some values, and it returns a list of nulls instead, this may mean that you misspelled the name of the field you intended to return, or used the wrong case (since field-names are case sensitive).

Operator Expressions

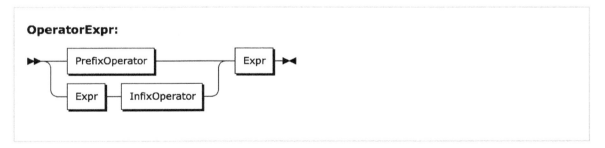

OperatorExpr:

SQL supports many operators that can be used before an expression or between two expressions. For the most part these operators also exist in N1QL and behave as you expect. But since N1QL supports objects and collections, we need to consider the effects of these new types on the operators of the language.

In general, if the operand(s) of an operator expression are of incompatible or inappropriate types, as in `5 + "Hello"`, the expression returns `null`.

Arithmetic operators: +, -, *, /, %(modulo), `div`, `mod`, ^(exponentiation)

> The + and − operators can be used in either infix (2 + 2) or prefix (-5) notation. The % and `mod` operators are synonyms for the same operation (modulo).

> Generally, if the result of an arithmetic operation can be accurately represented as an integer, it is returned as an integer; otherwise it is returned in decimal notation (Examples: `2.0 + 2.0` returns 4, but `2.0 + 2.1` returns `4.1`). An exception is the `div` operator, which returns an integer if both operands are integers (Example: `5 div 2` returns 2, but `5 div 2.0` returns `2.5`).

String concatenation operator: | |

> Example: `'Small' || 'talk'` returns `'Smalltalk'`.

Logical operators: AND, OR, NOT

> AND and OR are infix operators (take two operands), and NOT is a prefix operator (takes a single operand).

> When their operands are known, the logical operators follow the normal rules of logic. However, in the presence of special values, these operators behave as follows:

> - AND: If either operand is `false`, the result is `false`; otherwise if either operand is `missing`, the result is `missing`; otherwise if either operand is `null`, the result is `null`; otherwise the result is `true`.
>
> - OR: If either operand is `true`, the result is `true`; otherwise if either operand is `null`, the result is `null`; otherwise if either operand is `missing`, the result is `missing`; otherwise the result is `false`.
>
> - NOT: If the operand is `missing`, the result is `missing`; otherwise if the operand is `null`, the result is `null`; otherwise if the operand is `true`, the result is `false`; otherwise the result is `true`.

Truth Test Expressions

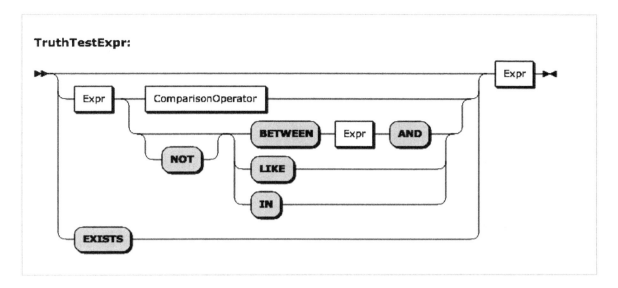

TruthTestExpr:

Truth test expressions are used wherever a `true` or `false` value is needed, as in a WHERE or HAVING clause.

The first thing to note is that any expression that returns `true` or `false` is a truth test expression. For example, the path expression `person.married`, or the index expression `raining[i]`, might return a truth value, and could appear in a WHERE or HAVING clause.

A very common kind of truth test expression is an expression that compares two values. Several kinds of comparison expressions are listed below. As a general rule, if an expression is comparing two dissimilar types, as in `5 > "Hello"`, or an operand does not have an appropriate type, as in `5 LIKE 27` or `5 IN 27`, the expression returns `null`.

Comparison operators: <, <=, >, >=, =, !=, <>

> Comparisons of two numbers or two strings apply as in SQL. As in the general rule, if one side of a comparison is `missing`, the comparison returns `missing`; otherwise if one side is `null`, the comparison returns `null`.

> Two objects are equal if they are "deeply equal"—that is, if they have the same named

fields (not necessarily in the same order) and the fields of the same name have equal values, using the same definition of equality for nested objects.

For objects, the < and > operators are defined by a system-defined stable ordering. In other words, the expression obj1 < obj2 will always return the same result, and if obj1 < obj2 is true, then obj1 > obj2 is false. The same stable ordering is used when ORDER BY is applied to objects. Since two objects might not contain comparable fields, the < and > operators for objects convey no semantic meaning.

Two arrays are equal if they have the same length and their sequences of items are pairwise equal (again, requiring "deep equality" for nested objects).

The comparison operators > and < are defined for arrays as follows: The items in the arrays are compared pairwise, moving from the beginning of the array to the end, until some pair of unequal values is found. That pair of values determines which array is larger. If one array is shorter than the other and the arrays are otherwise pairwise equal, the shorter array is considered to be smaller. Thus:

```
[1, 2, 3] < [1, 2, 4] is true
[1, 2] < [1, 2, 0] is true
[null, 99] < [0, 0] is null
```

When comparing two multisets, or comparing a multiset to an array, the multiset(s) are first converted to arrays by a system-defined process, and then the arrays are compared. The conversion process is stable (that is, two instances of the same multiset will convert into two instances of the same array).

Of course, it is a general rule, independent of types, that != is true if and only if = is false. Also <= is true if either < is true or = is true, and a similar rule applies to >=.

BETWEEN:

The expression: a BETWEEN b AND c
is equivalent to the expression: a >= b AND a <= c.

LIKE:

> The first and second operands of a LIKE expression must evaluate to strings; the second of these strings may contain the special characters '%' (representing zero or more characters) and/or '_' (representing a single character). For example, `'abcdefg'` `LIKE` `'a_cd%'` returns `true`. Both operands of LIKE may be any expression that evaluates to a string.

IN:

> The IN operator x `IN` y returns `true` if the value x is equal (by the = operator) to at least one value in the collection y. For example, the expression `5 IN [4, 5, 6]` returns `true`, and 5 `IN [ ]` returns `false`.

EXISTS:

> EXISTS is a prefix operator that returns `true` if its operand is a non-empty collection, and `false` if the operand is an empty collection. If the operand is not a collection, the EXISTS operator returns `null`. Like IN, the EXISTS operator is often used for linking to subqueries, but its use is not limited to subquery linkages. For example, `EXISTS [1]` is `true`, and `EXISTS [ ]` is `false`.

>ALL, >SOME, = ALL, =SOME, <ALL, <SOME, etc.:

> These SQL subquery linkage operators are not supported in N1QL, because their functionality is provided in a more general way by quantified expressions.

Function Calls

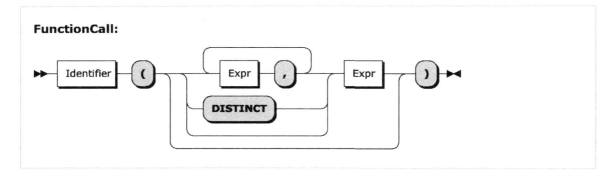

FunctionCall:

As in SQL, a function call consists of a function name followed by zero or more argument expressions enclosed in parentheses. N1QL supports a large library of functions (see Couchbase documentation for details).

The same function call syntax is used for the built-in SQL aggregation functions (SUM, AVG, MAX, MIN, and COUNT) and for the more conventional functions in the N1QL library (LENGTH, SUBSTR, etc.). However, the DISTINCT keyword may be used only with the built-in SQL aggregation functions.

Path Expressions

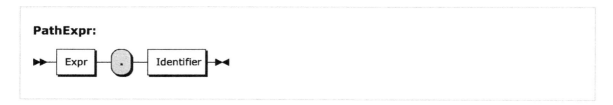

PathExpr:

Path expressions are a new kind of expression, not a part of SQL. A path expression consists simply of an expression that evaluates to an object, followed by a dot and an identifier, as in this example:

```
customer.address
```

In this example, `customer` is a variable that evaluates to an object, and the identifier `address` is used to find a field in that object whose name matches that identifier; that field is returned by the path expression.

It's easy to see that, if the result of a path expression is another object, it can serve as one "step" in a longer path expression, as in this example:

```
customer.address.zipcode
```

Note that the dot in a path expression is followed by an identifier, not by a string. For example, supposing that `customer` evaluates to an object, the expression `customer."address"` would be a syntax error, even if the `customer` object has a field whose name is `address`. Note also that, if the field-name in question is a reserved word or contains a special character, the identifier in the path expression must be enclosed in back-ticks, like this:

```
company.`union`
```

In this example, the identifier `` `union` `` is enclosed in back-ticks because it matches a reserved word.

We say that a path expression returns a field of an object. In most cases, this means that the expression returns the value of that field. For example, the expression `customer.address.zipcode` might return the string `"94040"` if that is the value of the `zipcode` field. The expression `customer.address.zipcode = "94040"` is a comparison of two strings. In certain contexts such as a SELECT clause, the name of the returned field is also preserved. For example, the clause `SELECT customer.address.zipcode` might return an object with a named field: `{"zipcode": "94040"}`. We'll learn more about SELECT clauses when we talk about queries.

In a path expression like `aaa.bbb`, if the object `aaa` does not contain a field named `bbb`, the path expression returns `missing` (you'll also get `missing` if you've misspelled the name of the field you're looking for, so be careful with your spelling. Also remember that field-names are case sensitive).

Index Expressions

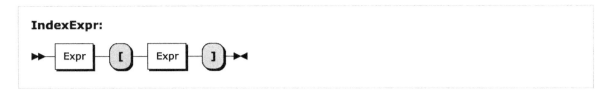

IndexExpr:

Index expressions, like path expressions, are an extension to SQL. An index expression is an expression that extracts an item from a collection. A simple example of an index expression is `a[i]`, where a represents an array and i represents an integer (of course, a could be any expression that returns an array, and i could be any expression that returns an integer). In this case, `a[i]` returns the ith item in the array a, using zero-based indexing, as shown in these examples:

```
a = ['Able', 'Baker', 'Charlie', 'Dog']
a[0] returns 'Able'
a[i + j] returns 'Dog' if i = 1 and j = 2
```

In the expression `a[i]`, if a is not a collection or i is not an integer, the expression returns `null`. If a is a collection and i is an integer but the item `a[i]` does not exist, the special value `missing` is returned.

Index expressions are usually used with arrays because arrays have a well-defined order. The items in a multiset, on the other hand, are not ordered. Therefore, in the expression `a[i]`, if a is a multiset and the index i is within the size of the multiset, the expression returns a random item in the multiset. Selection of the item is stable; that is, `a[i]` always returns the same item for the same index.

Of course, all expressions, including path and index expressions, can be nested inside each other to build queries that are as complex as you like. For example, suppose that a represents an array of objects that contain arrays of objects. Then the expression `a[3].b[5].c` means: "Find item 3 in a; then find the b-field in that item; then find item 5 in that field-value; then find the c-field in that item." As usual, when constructing complex expressions, you can use parentheses to specify the precedence of the operators.

Quantified Expressions

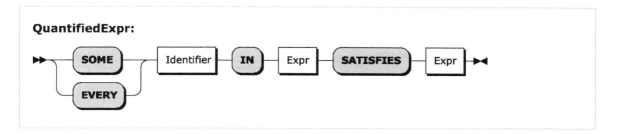

QuantifiedExpr:

N1QL introduces a new kind of expression called a *quantified expression*. Quantified expressions come in two forms, which start with the keywords SOME and EVERY. Apart from the initial keywords, both forms have the same syntax. We'll start by discussing the SOME expression, which looks like this:

```
SOME var IN expr1 SATISFIES expr2
```

The expression evaluates *expr1*, which must return a collection. Then it binds each item in that collection, in turn, to the variable *var*. For each of these bindings, it evaluates *expr2*, which is a Boolean expression that references *var*. If at least one of the evaluations of *expr2* returns true, the SOME expression returns true; otherwise, the SOME expression returns false. In a quantified expression, the keyword ANY has the same meaning as SOME.

SOME expressions can be easily understood by looking at a few examples:

- SOME x IN [1, 2, 3] SATISFIES x * 2 > 10

  This example evaluates the expression x * 2 > 10 three times, substituting the values 1, 2, and 3 for the variable x. All three evaluations return false:

  ```
  1 * 2 > 10  is false
  2 * 2 > 10  is false
  3 * 2 > 10  is false
  ```

Therefore the SOME expression returns `false`.

- `SOME x IN [4, 5, 6] SATISFIES x * 2 > 10`

    This example is similar to the previous one, but it returns `true` because one of the variable bindings results in a true expression: `6 * 2 > 10` is true.

The EVERY expression is similar to the SOME expression, except that it returns `true` only if *every* binding of *var* causes *expr2* to return `true`; otherwise it returns `false`. Again, we'll illustrate this expression through examples:

- `EVERY x IN [4, 5, 6] SATISFIES x * 2 > 10`

    This example returns `false` because the expression `x * 2 > 10` is true for some values of x but not for every value of x.

- `EVERY x IN [4, 5, 6] SATISFIES x * 2 > 5`

    This example returns `true` because the expression `x * 2 > 5` is true for every value of x in the set `[4, 5, 6]`.

- `EVERY x IN [ ] SATISFIES x * 2 > 5`

    This example returns `true`. An EVERY expression always returns `true` if the expression following `IN` returns an empty collection.

CASE Expressions

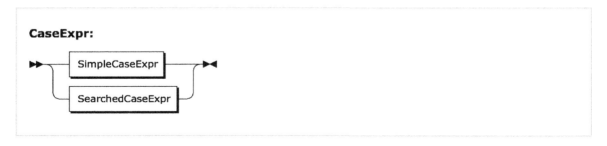

CaseExpr:

SimpleCaseExpr:

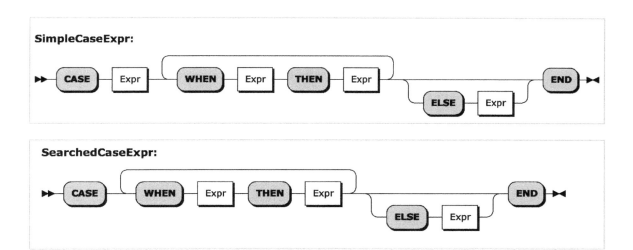

SearchedCaseExpr:

As in SQL, a CASE expression in N1QL has two forms. In a simple CASE expression, one of several output expressions is evaluated, depending on the value of an input expression. In a searched CASE expression, a list of test expressions is provided, each with its own output expression; the CASE expression evaluates the output expression corresponding to the first test expression that returns `true`. The semantics of CASE expressions are the same in SQL and N1QL.

Subqueries

Subquery:

In N1QL, a subquery may appear anywhere a value is expected. A subquery is simply a query wrapped in parentheses.

There is one important difference between SQL and N1QL in the way subqueries are handled. In SQL, if a subquery returns a single value (one row with one column), that result is treated as a scalar value. The following example is valid in SQL because the result of the subquery is treated as a number. If the average price of parts is 150, the comparison would return `true`.

```
... WHERE (SELECT AVG(price) FROM parts) > 100
```

In N1QL, on the other hand, a subquery always returns a collection, even if the collection is empty or contains a single value. N1QL does not automatically extract the value from the collection. Therefore, in the example above, the subquery would return a collection, and the comparison would fail (return `null`) because a collection is not comparable to a number. To compare the result of the subquery to a number, you must use an index expression to extract the number from the collection. The N1QL code fragment that is equivalent to the above example is as follows:

```
... WHERE (SELECT VALUE AVG(price) FROM parts)[0] > 100
```

We'll talk more about SELECT VALUE queries in a later section.

Temporal Data

As noted above, the only primitive values supported by JSON are strings, numbers, `true`, `false`, and `null`. However, temporal values such as dates and times are important to many applications. N1QL supports a large set of temporal functions that permit dates and times to be represented using either strings or numbers.

Temporal data can be represented by a string containing several components separated by delimiter characters, using a notation defined in the ISO 8601 standard. Here are some examples:

- `"2017-07-02T15:04:05.567-8:00"`
  This example includes all the components that apply to temporal data: year, month, day, hour, minute, second, and timezone. A fully-populated temporal string like this one is called a timestamp. This timestamp represents July 2, 2017, at 4 minutes and 5.567 seconds after 3 p.m., in a timezone that (like San Francisco) is eight hours behind Coordinated Universal Time (UTC). If no timezone is specified, the local timezone at your Couchbase server is assumed. If any other component is omitted, it defaults to the value zero.

- `"2017-07-02"`
  This example shows how a date can be represented by including year, month, and day, and omitting other components.

- "08:15"

  This example shows how a time (in this case, 8:15 a.m.) can be represented by including hour and minute and omitting other components. The "T" character is used when a temporal value includes both a date and a time, such as "2017-07-02T08:15".

Temporal data can also be represented by a number that represents an integer number of milliseconds after midnight on January 1, 1970 UTC (sometimes called Epoch/Unix time). N1QL functions generally accept temporal data in either string or numeric format. For example, the DATE_DIFF_STR function finds the difference between two dates represented as strings, and the DATE_DIFF_MILLIS function finds the difference between two dates represented in numeric (milliseconds) format.

N1QL supports a large set of functions for processing temporal data. The following examples illustrate two of these functions.

Suppose date1 is the string "2017-06-15" and date2 is the string "2017-07-20". Then:

```
DATE_DIFF_STR(date2, date1, "day") = 35
```

35 is the number of days between date1 and date2. The third parameter of DATE_DIFF_STR ("day" in this example) identifies the component to be returned. If the third parameter had been "month", the function would have returned the number 1.

The function DATE_PART_STR returns one of the components of a date represented as a string. Although the date is represented as a string, the selected component is returned as a number. Using the values of date1 and date2 in the previous example:

```
DATE_PART_STR(date1, "year") = 2017
DATE_PART_STR(date2, "month") = 7
```

A complete list of temporal functions can be found in the N1QL documentation.

Errors

In N1QL, as in other languages, there are many ways to write a query that is syntactically valid but doesn't make any sense. For example, you can use the dot operator on something that's not an object, or you can use an array index on something that's not an array, or you can try to add a number to a string. It's not generally possible in N1QL to detect errors like this before processing a query, since the compiler often doesn't have schema information that allows it to compute the type of each expression.

The general rule in N1QL, when an expression is encountered that is syntactically valid but makes no sense, is simply to return null and to proceed with evaluating the rest of the query. This rule enables the system to return results when they are meaningful and to ignore expressions that are not meaningful.

As a result of this general rule, you probably won't see as many error messages from N1QL as you are used to seeing with SQL. In general, if you ask a question that doesn't make any sense, you just won't get any answer. The downside of this approach is that errors may be hard to catch, because a query containing an error may run to completion and return a null value, an empty collection, or a result that does not include some of the expected data.

Queries

Queries

We're now ready to look at some actual queries. We'll begin with queries that look familiar to you, and then discuss a new type of query called SELECT VALUE. We'll discuss some special considerations that apply to subqueries in N1QL, and then we'll dive into the details of each of the clauses in a query block.

One note before we begin: as in SQL, keywords in N1QL are not case sensitive. By convention, in this tutorial we will use uppercase keywords to distinguish them from variable names. All N1QL keywords are reserved words, so if you want to reference a dataset or field whose name is the same as a keyword, you'll need to enclose the name in back-ticks `like this`.

A comment on comments: N1QL allows two kinds of comments in queries—block comments and line comments. Block comments extend from the beginning delimiter /* to the ending delimiter */. Line comments extend from the beginning delimiter // or -- to the end of the line. Here's what they look like:

```
Block comments    /* LIKE THIS */
Line comments     -- LIKE THIS
Line comments     // LIKE THIS
```

Sample Data

We'll discuss many of the features of N1QL by running some example queries based on two simple datasets called `customers` and `orders`. The `customers` dataset contains a multiset of objects, each of which has a customer id, a name, an address, and a credit rating. The `orders` dataset contains a multiset of objects, each of which has an order number, order and ship dates, a customer id, and an array of items. Since these datasets have a regular structure, you might think of them as tables, and you might think of `custid` as the primary key of the `customers` table and as a foreign key in the `orders` table. However, unlike tables in the relational model, these datasets contain nested objects and nested arrays. Also unlike relational tables, some of the objects inside `customers` and `orders` may have missing fields or additional fields.

A typical object in the `customers` dataset looks like this:

```
{ "custid": "C13",
```

```
        "name": "T. Cruise",
        "address":
           { "street": "201 Main St.",
             "city": "St. Louis, MO",
             "zipcode": "63101"
           },
        "rating": 750
     }
```

A typical object in the `orders` dataset looks like this:

```
{ "orderno": 1007,
  "custid": "C13",
  "order_date": "2017-09-13",
  "ship_date": "2017-09-20",
  "items": [ { "itemno": 185,
               "qty": 5,
               "price": 21.99
             },
             { "itemo": 680,
               "qty": 1,
               "price": 20.50
             }
           ]
}
```

I've deliberately kept the sample database small so that you can see all the data and the results of the example queries. The complete database is shown in Appendix A at the back of the book.

SELECT Queries

As we've seen, the objects in a dataset can be considered as analogous to rows of a table. We'll begin our discussion with some queries on the `customers` dataset, focusing on customers' zipcodes and credit ratings. To get started, we'll display all the zipcode and rating information in the dataset, so you can see it all together. The following query should look familiar to you. Note that since zipcode data is nested inside an address object, we need to refer to it as `address.zipcode`. Note also that customer C31 has no rating and customer C47 has no zipcode.

(Q1) List the customer id, name, zipcode, and credit rating of all customers, in order by customer id.

```
SELECT custid, name, address.zipcode, rating
FROM customers
ORDER BY custid;
```

Result:

```
[ { "custid": "C13",
    "name": "T. Cruise",
    "zipcode": "63101"
    "rating": 750,
  },
  { "custid": "C25",
    "name": "M. Streep",
    "zipcode": "02340"
    "rating": 690,
  },
  { "custid": "C31",
    "name": "B. Pitt",
    "zipcode": "63101"
  },
  { "custid": "C35",
    "name": "J. Roberts",
    "zipcode": "02115"
    "rating": 565,
  },
  { "custid": "C37",
    "name": "T. Hanks",
    "zipcode": "02115"
    "rating": 750,
  },
  { "custid": "C41",
    "name": "R. Duvall",
    "zipcode": "63101"
    "rating": 640,
  },
```

```
  { "custid": "C47",
    "name": "S. Loren",
    "rating": 625
  }
]
```

As you know, SQL allows you to assign a variable-name to each collection in the FROM clause. In this tutorial, I'll refer to these variable-names as *iteration variables*. As a query is executed, an iteration variable is bound, in turn, to each item in the associated collection. For example, the clause FROM customers AS c iterates over the customers dataset, binding the iteration variable c to each customer object in turn. The iteration variable can then be used elsewhere in the query to refer to the fields of the object bound to that variable. For example, c.name refers to the name field of the customer object bound to the variable c. Note that the name customer denotes the dataset as a whole, and the name c denotes one object inside the dataset.

In example query Q2, we define the iteration variable c in the FROM clause and use it to qualify field-names in the SELECT and WHERE clauses.

(Q2) Find the name and rating of the customer with id C41.

```
SELECT c.name, c.rating
FROM customers AS c
WHERE c.custid = "C41";
```

Result:

```
[ { "name": "R. Duvall",
    "rating": 640
  }
]
```

Iteration variables aren't always required, but they are helpful in understanding how a query works. And some cases, such as when joining a dataset to itself, they are actually necessary. We'll discuss iteration variables further in a later part of this tutorial.

As we've seen, the job of a FROM clause is to create a stream of variable bindings (in Q2, it binds the variable c to the objects in the customers dataset, one at a time). If no iteration variable is specified in the FROM clause, an implicit iteration variable is created that has the same name as the dataset (that's why Q1 worked without an iteration variable). Some of the bindings may be filtered out by the WHERE clause. The SELECT clause then creates an output object from each surviving binding. The result is always a collection containing zero or more objects. If the query has an ORDER BY clause, the result is an array; otherwise it is a multiset. As mentioned earlier, when a query result is serialized into JSON format, a multiset is represented as an JSON array with arbitrary order.

SELECT VALUE Queries

We're now ready to talk about a new kind of query called a SELECT VALUE query. Like a SELECT query, a SELECT VALUE query returns a collection of results, but the items in the collection are not restricted to objects; in fact they can be any value in the JSON data model.

The simplest form of a SELECT VALUE query consists simply of the words SELECT VALUE followed by an expression. The query evaluates the expression and returns the result, wrapped in an array, as in this example:

```
SELECT VALUE 2 + 2;
```

Result: [4]

In this simple form, SELECT VALUE must always be followed by a single expression. The following example is an error because it is not a valid single expression:

```
SELECT VALUE 1, 2, 3;
```

The expression following SELECT VALUE can be any N1QL expression, including an expression that returns a complex result such as an array, an object, or an array of objects. In the following example, the expression following SELECT VALUE returns an array. As always, N1QL wraps the result in another array:

```
SELECT VALUE [1, 2, 3];
```

Result:

```
[ [ 1, 2, 3 ] ]
```

A more complex form of a SELECT VALUE query may contain a FROM clause; in fact it may contain all the clauses in a traditional SQL query. If a SELECT VALUE query has a FROM clause, it differs from an ordinary SELECT query in the following ways:

- The SELECT VALUE clause must contain only a single expression, with no alias.

- The result of a SELECT VALUE query is simply a collection of values for the selected expression, one for each binding generated by the FROM clause. Unlike an ordinary SELECT query, the SELECT VALUE query does not give names to the resulting values or wrap them in an object. If ORDER BY is specified, the result is an array; otherwise it is a multiset.

The differences between SELECT and SELECT VALUE are illustrated by the following examples:

(Q3) Find the names of customers with a rating greater than 650.

```
SELECT name
FROM customers
WHERE rating > 650;
```

Result:

```
[ { "name": "T. Cruise" },
  { "name": "M. Streep" },
  { "name": "T. Hanks" }
]
```

(Q4) Shows the effects of SELECT VALUE (compare to Q3).

```
SELECT VALUE name
FROM customers
```

```
WHERE rating > 650;
```

Result:

```
[ "T. Cruise", "M. Streep", "T. Hanks" ]
```

(Q5) Returns a multiset of [name, rating] pairs for customers with rating greater than 650.

```
SELECT VALUE [name, rating]
FROM customers
WHERE rating > 650;
```

Result:

```
[
  [ "T. Cruise", 750 ],
  [ "M. Streep", 690 ],
  [ "T. Hanks", 750 ]
]
```

Note that the result of a SELECT VALUE query, although it may be derived from fields that have names, does not include any of these field names.

A SELECT VALUE query can be used with an object constructor to create labels or to give some structure to a query result, as in the following example.

(Q6) List customers with credit rating greater than 650, in order by descending credit rating, and again in ascending order by zipcode.

```
SELECT VALUE
  {"high-rated customers, ordered by rating":
      (SELECT c.rating, c.custid, c.name
       FROM customers AS c
       WHERE c.rating > 650
       ORDER BY c.rating DESC),
  "high-rated customers, ordered by zipcode":
```

```
        (SELECT c.address.zipcode, c.custid, c.name
         FROM customers AS c
         WHERE c.rating > 650
         ORDER BY c.address.zipcode)
    };
```

Result:

```
[ { "high-rated customers, ordered by rating":
      [ { "rating": 750,
          "custid": "C13",
          "name": "T. Cruise"
        },
        { "rating": 750,
          "custid": "C37",
          "name": "T. Hanks"
        },
        { "rating": 690,
          "custid": "C25",
          "name": "M. Streep"
        }
      ],
    "high-rated customers, ordered by zipcode":
      [ { "zipcode": "02115",
          "custid": "C37",
          "name": "T. Hanks"
        },
        { "zipcode": "02340",
          "custid": "C25",
          "name": "M. Streep"
        },
      { "zipcode": "63101",
        "custid": "C13",
        "name": "T. Cruise"
      }
    ]
  }
]
```

For historical reasons, the keywords ELEMENT or RAW can be used in place of VALUE in a SELECT VALUE query. In this tutorial, we will consistently use the word VALUE.

Subqueries

A *subquery* is simply a query that is nested inside another query, enclosed in parentheses. N1QL allows a subquery to be used wherever a value is expected. For example, a subquery might be used in a FROM clause; the FROM clause would then iterate over the items returned by the subquery. Similarly, if a subquery appears in the SELECT clause, the result of that subquery is included in the output of the main query.

Subqueries give you a great deal of power to write complex queries. Along with this power comes a necessity to write your queries very carefully. We'll illustrate both the power and the necessity for caution with a series of examples. These examples may be helpful in diagnosing problems that you might encounter in writing subqueries of your own.

Suppose, for a promotional mailing, you need a list of the customers who have the highest credit rating (there may be more than one in the case of a tie). Since you are an experienced SQL user, you can easily imagine how this query might be written in SQL using a subquery like this:

(Q7) Find the name(s) of the customers who have the highest credit rating.

```
SELECT name
FROM customers
WHERE rating =
  (SELECT MAX(rating) FROM customers);
```

Based on your SQL experience, you might expect the result of Q7 to be a simple list of names, perhaps something like this:

```
[ "T. Cruise", "T. Hanks" ]
```

Instead, when you try to run Q7, you get an error message. The message isn't very helpful, but here's what's wrong: you have referenced the customers dataset twice in this query without using any iteration variables. The compiler is confused, and is interpreting customers in the subquery as the single customer bound by the FROM clause of the outer query.

Q7 would have been a valid query in SQL, but N1QL, which has no schema information to rely on, needs some extra help. The solution is to introduce two iteration variables to distinguish the two references to customers, giving us this version of the query:

(Q8) Same as Q7, with iteration variables to resolve ambiguities.

```
SELECT c1.name
FROM customers AS c1
WHERE c1.rating =
    (SELECT MAX(c2.rating)
     FROM customers AS c2);
```

Result: [ ]

Once again, you might expect this query to return two names: T. Cruise and T. Hanks. That is what this query would do in SQL. However, in N1QL, the query runs with no error messages and returns an empty result. Now you're wondering how it can be that no customer has a credit rating equal to the maximum credit rating of any customer. You don't have an error message to help you diagnose this problem.

As you study the problem, you'll remember that, in N1QL, every SELECT query returns a collection (array or multiset) of objects, regardless of whether it's a subquery or not. With this in mind, let's look at the linkage between the outer query and the subquery. The linkage is made by an "=" operator, which wants to compare two values. The value on the left side of the "=" operator is c1.rating, which is a number like 750. The value on the right side of the "=" operator is the result of the subquery. The subquery is a SELECT query and it returns a collection of objects. Clearly a number like 750 is not comparable to a collection of objects, so the "=" comparison returns null and the query returns an empty result.

The first step in fixing this problem is to change the subquery from SELECT to SELECT VALUE. This will cause the subquery to return a collection of values that are not wrapped inside objects. (Of course, there's only one value in the collection, but it's still a collection.)

The second step is to change the way in which the results of the subquery are compared with the value of the field c1.rating. The basic problem is that c1.rating is a number, but the subquery returns a collection containing one number. In cases like this, SQL would automatically extract the number from the collection and perform the comparison. N1QL, in contrast, does not perform this automatic extraction. Instead, it carefully distinguishes between a scalar value and a collection of values, and does not automatically convert one into the other. This is one of the main incompatibilities between SQL and N1QL, and it often leads to a problem when a subquery is compared to a scalar value. We'll refer to this problem as the "subquery problem."

One way to solve the subquery problem is simply to change the subquery linkage operator (the operator that connects the subquery to the outer query) from "=" to IN. The IN operator compares a number like 750 to a collection of numbers. It returns true if the number is equal to any number in the collection. Using IN as a comparison operator, our query can be expressed as follows:

(Q9) Still trying to find the name(s) of customers with the highest credit rating.

```
SELECT c1.name
FROM customers AS c1
WHERE c1.rating IN
    (SELECT VALUE MAX(c2.rating) FROM customers AS c2);
```

Result:

```
[ { "name": "T. Cruise" },
  { "name": "T. Hanks" }
]
```

This solution to the subquery problem has one limitation: it works only when the subquery linkage operator is "=". If the subquery linkage operator had been ">", converting it to IN would not be an option. For example, suppose that rather than the names of the customers with the highest rating, we only want the names of customers whose rating is higher than the average rating.

One way to solve this problem is to use the index operator [0] to extract a value from the collection of values returned by the subquery. In our case there's only one value in the collection.

This approach works well for all kinds of subquery linkages.

> (Q10) List the names of the customers whose credit rating is higher than the average rating.

```
SELECT c1.name
FROM customers AS c1
WHERE c1.rating >
    (SELECT VALUE AVG(c2.rating) FROM customers AS c2)[0];
```

> Result:

```
[ { "name": "T. Cruise" },
  { "name": "M. Streep" },
  { "name": "T. Hanks" }
]
```

If you would like the result of Q7 to be simply a list of names, not wrapped in objects, you can use SELECT VALUE in the outer query as well:

> (Q11) Find the names of the customer(s) who have the highest credit rating, as an unlabeled collection of names.

```
SELECT VALUE c1.name
FROM customers AS c1
WHERE c1.rating =
    (SELECT VALUE MAX(c2.rating) FROM customers AS c2)[0];
```

> Result:

```
[ "T. Cruise", "T. Hanks" ]
```

This result finally looks like the one that you might have expected to get from Q7.

In the Expressions section, we introduced a feature called a quantified expression. You might be thinking that a quantified expression could be used as an alternative way to find the highest-rated customers. Let's try it and see what happens.

(Q12) Find the names of the customer(s) who have the highest credit rating, using a quantified expression.

```
SELECT VALUE c1.name
FROM customers AS c1
WHERE EVERY r IN
    (SELECT VALUE c2.rating FROM customers AS c2)
SATISFIES c1.rating >= r;
```

Result: [ ]

Once again, we have a query that runs with no error messages and returns an unexpected empty result. This time, the culprit is the object in the `customers` dataset that has no `rating` field. The query searches for customers whose rating is greater than or equal to the rating of every other customer. But customer C31 has no rating, and so any attempt to compare some rating with the rating of customer C31 returns the value `missing`, which in turn causes the EVERY predicate to return `false`. As a result, no customer satisfies the EVERY predicate and the query returns an empty result.

The way to fix this problem is to insert an additional test into the subquery: `WHERE c1.rating IS KNOWN`. This test will cause the EVERY predicate to apply only to customers whose revenue is known (that is, not `null` and not `missing`), and the query will return the expected results, as shown here:

(Q13) Same as Q12, with a provision to handle null and missing values.

```
SELECT VALUE c1.name
FROM customers AS c1
WHERE EVERY r IN
```

```
       (SELECT VALUE c2.rating
          FROM customers AS c2
         WHERE c2.rating IS KNOWN)
    SATISFIES c1.rating >= r;
```

Result:

```
[ "T. Cruise", "T. Hanks" ]
```

To summarize the lessons of this section: subqueries can be tricky. To avoid ambiguities, it's wise to define iteration variables in your FROM clauses and to use them in your field references. It's also wise to use SELECT VALUE in your subquery, and to use the IN operator as your subquery linkage or to use the indexing operator [0] to extract a scalar value. Also, if you're using a quantified expression, don't forget to add a test for IS KNOWN.

Query Blocks

Query Blocks

We'll use the term *query block* to denote the basic building block of a query: a collection of clauses that may include SELECT, FROM, LET, WHERE, GROUP BY, HAVING, ORDER BY, LIMIT, and OFFSET. For reasons that will soon be apparent, we'll refer to the FROM, LET, WHERE, GROUP BY, and HAVING clauses as the "Stream Generator," and the ORDER BY, LIMIT, and OFFSET clauses as "Output Modifiers."

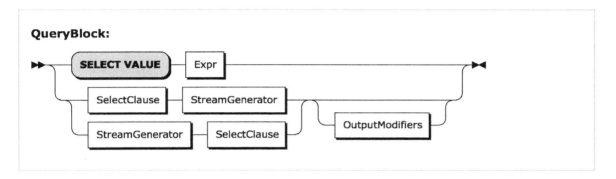

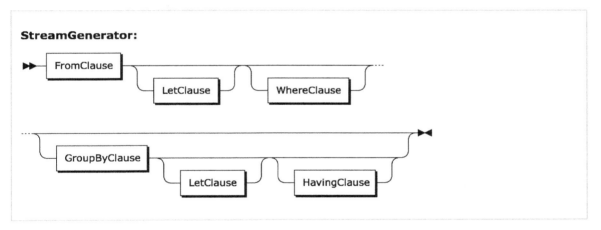

As an SQL user, you're familiar with the basic structure and purpose of a query block. The FROM clause iterates over an input dataset and binds a variable to each of its contents, in turn. (In SQL, you can think of these variable bindings as rows of a table; in N1QL the bindings can be more heterogeneous but I'll still use the term "row" here because it's a useful analogy.) The optional WHERE clause applies a test to each row and may filter some of them out. The optional GROUP BY clause, if present, forms the rows into groups and limits you to examining the properties of groups rather than individual rows. The optional HAVING clause applies a test to each group and may filter some of them out (of course, HAVING makes sense only if GROUP BY is present). The result of all these clauses (FROM, WHERE, GROUP BY, and HAVING) is a stream of variable bindings (you can think of them as rows). For this reason, I've called this group of clauses a "StreamGenerator" in the syntax diagram.

For each "row" (set of variable bindings) in the stream, the SELECT clause constructs an output object. The optional ORDER BY, LIMIT, and OFFSET clauses impose an ordering on the output objects and may place an upper limit on their number or skip some of the initial objects (of course, OFFSET makes sense only if ORDER BY is present). We'll discuss ORDER BY, LIMIT, and OFFSET clauses later.

So far, our description of a N1QL query block looks pretty much like an SQL query block. In addition, N1QL relaxes some constraints and provides some new query features. We'll first look at some of the new features that apply to the query block as a whole, and then we'll discuss each clause in turn.

One of the first things you may notice about a N1QL query block is that, unlike a query block in SQL, it is not required to have a FROM clause. In N1QL, a query block may simply consist of SELECT VALUE followed by a single expression. In that case, the semantics of the block are simple: it simply returns a multiset containing the value of the expression. For example, SELECT VALUE 7 returns [7].

In a query block that has no FROM clause, you're allowed to write either SELECT VALUE or SELECT DISTINCT VALUE. I've left DISTINCT out of the syntax diagram because it has no effect in this case. If there's no FROM clause, the query block is not doing any iteration; it is simply evaluating a single expression and returning its value, wrapped in a multiset, so there are no duplicates to eliminate (even if the single value happens to be an array).

Another thing you may notice about a N1QL query block is that, when a FROM clause is present, the SELECT clause is allowed to come either first (before the "StreamGenerator" clauses) or later (after the "StreamGenerator" clauses). The latter form is sometimes easier to understand because it corresponds more closely with the way queries are actually processed (as noted above, the SELECT clause is not applied until groups have been formed by the GROUP BY clause and filtered by the HAVING clause). In our tutorial examples, we'll sometimes take advantage of this new flexibility by writing the SELECT clause in the "later" position.

You may also notice that N1QL introduces a new optional clause called LET, which follows the FROM clause. This clause allows you to bind one or more variables to any values you like, which sometimes helps you to avoid typing the same expression multiple times. We'll talk more about the LET clause in later sections.

FROM Clause

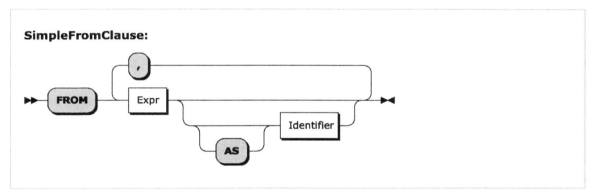

As we've discussed earlier, the purpose of a FROM clause is to iterate over a collection, binding a variable to each item in turn. Here's a query that iterates over the customers dataset, choosing certain customers and returning some of their attributes. This query should be so familiar to you as an SQL user that it needs no explanation.

> (Q14) List the customer ids and names of the customers in zipcode 63101, in order by their customer ids.

```
FROM customers
WHERE address.zipcode = "63101"
SELECT custid AS customer_id, name
ORDER BY customer_id;
```

Result:

```
[ { "customer_id": "C13",
    "name": "T. Cruise"
  },
  { "customer_id": "C31",
    "name": "B. Pitt"
  },
  { "customer_id": "C41",
    "name": "R. Duvall"
  }
]
```

Let's take a closer look at what this FROM clause is doing. A FROM clause always produces a stream of bindings, in which an iteration variable is bound in turn to each item in a collection. In Q14, since no explicit iteration variable is provided, the FROM clause defines an implicit variable named customers, the same name as the dataset that is being iterated over. The implicit iteration variable serves as the object-name for all field-names in the query block that do not have explicit object-names. Thus, address.zipcode really means customers.address.zipcode, custid really means customers.custid, and name really means customers.name.

You may also provide an explicit iteration variable, as in this version of the same query:

(Q15) Alternative version of Q14 (same result).

```
FROM customers AS c
WHERE c.address.zipcode = "63101"
SELECT c.custid AS customer_id, c.name
ORDER BY customer_id;
```

In Q15, the variable c is bound to each customer object in turn as the query iterates over the customers dataset. An explicit iteration variable can be used to identify the fields of the referenced object, as in c.name in the SELECT clause of Q15. When referencing a field of an object, the iteration variable can be omitted when there is no ambiguity. For example, c.name could be replaced by name in the SELECT clause of Q15. That's why field-names like name and custid could stand by themselves in the Q14 version of this query.

In the examples above, the FROM clause iterates over the objects in a dataset. But in general, a FROM clause can iterate over any collection. For example, the objects in the orders dataset each contain a field called items, which is an array of nested objects. In some cases, you will write a FROM clause that iterates over a nested array like items.

The stream of objects (more accurately, variable bindings) that is produced by the FROM clause does not have any particular order. The system will choose the most efficient order for the iteration. If you want your query result to have a specific order, you must use an ORDER BY clause.

In this tutorial, I will almost always specify an explicit iteration variable for each collection in the FROM clauses of my queries, and I will use these variables to qualify the field-names in other clauses. I recommend that you do the same when you are writing N1QL. I think explicit iteration variables are good practice for the following reasons:

- It's nice to have different names for the collection as a whole and an object in the collection. For example, in the clause FROM customers AS c, the name customers represents the dataset and the name c represents one object in the dataset.

- In some cases, iteration variables are required. For example, when joining a dataset to itself, distinct iteration variables are required to distinguish the left side of the join from the right side.

- In a subquery it's sometimes necessary to refer to an object in an outer query block (this is called a *correlated subquery*). To avoid confusion in correlated subqueries, it's best to use explicit variables. (We saw an example of this problem in Q7.)

Joins

A FROM clause gets more interesting when there is more than one collection involved. The following query iterates over two collections: customers and orders. The FROM clause produces a stream of *binding tuples*, each containing two variables, c and o. In each binding tuple, c is bound to an object from customers, and o is bound to an object from orders.

Conceptually, at this point, the binding tuple stream contains all possible pairs of a customer and an order (this is called the *Cartesian product* of `customers` and `orders`). Of course, we are interested only in pairs where the `custid` fields match, and that condition is expressed in the WHERE clause, along with the restriction that the order number must be 1001.

(Q16) Create a packing list for order number 1001, showing the customer name and address and all the items in the order.

```
FROM customers AS c, orders AS o
WHERE c.custid = o.custid
AND o.orderno = 1001
SELECT o.orderno,
       c.name AS customer_name,
       c.address,
       o.items AS items_ordered;
```

Result:

```
[ { "orderno": 1001,
    "customer_name": "R. Duvall",
    "address":
        { "city": "St. Louis, MO",
          "street": "150 Market St.",
          "zipcode": "63101"
        },
    "items_ordered":
        [ { "itemno": 347,
            "price": 19.99,
            "qty: 5
          },
          { "itemno": 193,
            "price": 28.89,
            "qty": 2
          }
        ]
    }
]
```

As you know, Q16 is called a *join query* because it joins the `customers` collection and the `orders` collection, using the join condition `c.custid = o.custid`. In N1QL, as in SQL, you can express this query more explicitly by a JOIN clause that includes the join condition, as follows:

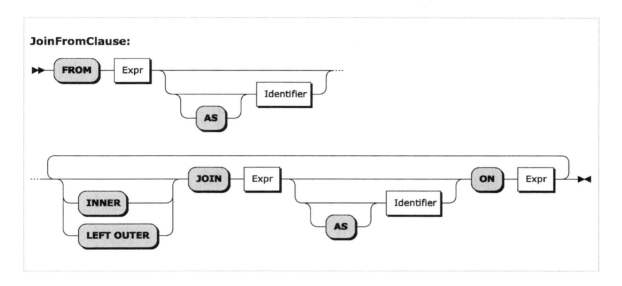

JoinFromClause:

(Q17) Alternative statement of Q16 (same result).

```
FROM customers AS c JOIN orders AS o
   ON c.custid = o.custid
WHERE o.orderno = 1001
SELECT o.orderno,
       c.name AS customer_name,
       c.address,
       o.items AS items_ordered;
```

Whether you express the join condition in a JOIN clause or in a WHERE clause is a matter of taste; the result is the same. In the examples in this tutorial, I'll generally use a comma-separated list of collection-names in the FROM clause, leaving the join condition to be expressed elsewhere. As we'll soon see, in some query blocks the join condition can be omitted entirely.

There is, however, one case in which an explicit JOIN clause is necessary. That is when you need to join collection A to collection B, and you want to make sure that every item in collection A is present in the query result, even if it doesn't match any item in collection B. This kind of query is called a *left outer join*, and it is illustrated by the following example.

(Q18) List all customers by customer id and name, together with the order numbers and dates of their orders (if any).

```
FROM customers AS c
     LEFT OUTER JOIN orders AS o ON c.custid = o.custid
SELECT c.custid, c.name, o.orderno, o.order_date
ORDER BY c.custid, o.order_date;
```

Result:

```
[ { "custid": "C13",
    "name": "T. Cruise",
    "orderno": 1002,
    "order_date": "2017-05-01"
  },
  { "custid": "C13",
    "name": "T. Cruise",
    "orderno": 1007,
    "order_date": "2017-09-13"
  },
  { "custid": "C13",
    "name": "T. Cruise",
    "orderno": 1008,
    "order_date": "2017-10-13"
  },
  { "custid": "C25",
    "name": "M. Streep"
  },
  ...
]
```

As you can see from the result of this left outer join, our data includes three orders from customer T. Cruise, but no orders from customer M. Streep. The behavior of left outer join in N1QL is different from that of SQL. SQL would have provided M. Streep with an order in which all the fields were null. N1QL, on the other hand, deals with schemaless data, which permits it to simply omit the order fields from the outer join.

Now we're ready to look at a new kind of join that was not provided (or needed) in original SQL. Consider this query:

(Q19) For every case in which an item is ordered in a quantity greater than 100, show the order number, date, item number, and quantity.

```
FROM orders AS o, o.items AS i
WHERE i.qty > 100
SELECT o.orderno, o.order_date, i.itemno AS item_number,
        i.qty AS quantity
ORDER BY o.orderno, item_number;
```

Result:

```
[ { "orderno": 1002,
    "order_date": "2017-05-01",
    "item_number": 680,
    "quantity": 150
  },
  { "orderno": 1005,
    "order_date": "2017-08-30",
    "item_number": 347,
    "quantity": 120
  },
  { "orderno": 1006,
    "order_date": "2017-09-02",
    "item_number": 460,
    "quantity": 120
  }
]
```

Q19 illustrates a feature called *left-correlation* in the FROM clause. Notice that we are joining orders, which is a dataset, to items, which is an array nested inside each order. In effect, for each order, we are unnesting the items array and joining it to the order as though it were a separate collection. For this reason, this kind of query is sometimes called an *unnesting* query. The keyword UNNEST may be used whenever left-correlation is used in a FROM clause, as shown in this example:

(Q20) Alternative statement of Q19 (same result).

```
FROM orders AS o UNNEST o.items AS i
WHERE i.qty > 100
SELECT o.orderno, o.order_date, i.itemno AS item_number,
       i.qty AS quantity
ORDER BY o.orderno, item_number;
```

The results of Q19 and Q20 are exactly the same. In Q20, the word UNNEST does not affect the result of the query (and therefore I've omitted it from the syntax diagram). UNNEST simply serves as a reminder that left-correlation is being used to join an object with its nested items. (You can also use the noise-words CORRELATE or FLATTEN in place of UNNEST—choose your favorite!) The join condition in Q20 is expressed by the left-correlation: each order o is joined to its own items, referenced as o.items. The result of the FROM clause is a stream of binding tuples, each containing two variables, o and i. The variable o is bound to an order and the variable i is bound to one item inside that order.

Before ending our discussion of the FROM clause, let's consider the following query:

(Q21) Find the customer ids and names of all customers who ordered item no. 680, and the dates of their orders.

Like many queries, Q21 can be expressed in several different ways. Here are two of them:

(Q21a) Here Q21 is expressed as a three-way join with left-correlation.

```
FROM orders AS o, o.items AS i, customers AS c
WHERE o.custid = c.custid
```

```
AND i.itemno = 680
SELECT c.custid, c.name, o.order_date AS date
ORDER BY c.custid, date;
```

Result:

```
[ { "custid": "C13",
    "name": "T. Cruise",
    "date": "2017-05-01"
  },
  { "custid": "C13",
    "name": "T. Cruise",
    "date": "2017-09-13"
  },
  { "custid": "C35",
    "name": "J. Roberts",
    "date": "2017-07-10"
  },
  { "custid": "C41",
    "name": "R. Duvall",
    "date": "2017-09-02"
  }
]
```

(Q21b) Here Q21 is expressed as a two-way join with a correlated subquery, with the same result.

```
FROM orders AS o, customers AS c
WHERE o.custid = c.custid
AND EXISTS
    (SELECT i.itemno
      FROM o.items AS i
      WHERE i.itemno = 680)
SELECT c.custid, c.name, o.order_date AS date
ORDER BY c.custid, date;
```

LET Clause

LetClause:

The purpose of a LET clause, which immediately follows a FROM clause, is simply to bind one or more variables that can be referenced in other parts of the query block. Consider the following query:

> (Q22) Your company policy is to ship each order within two days. List all the late orders, in descending order by the number of days each order is late.

```
FROM orders AS o
LET days = DATE_DIFF_STR(o.ship_date, o.order_date, "day")
WHERE days > 2
SELECT o.orderno, days - 2 AS days_late
ORDER BY days_late DESC;
```

> Result:

```
[ { "orderno": 1007,
    "days_late": 5
  },
  { "orderno": 1004,
    "days_late": 3
  },
  { "orderno": 1001,
    "days_late": 2
  }
]
```

For each order bound by the FROM clause to the variable o, the LET clause assigns the name days to the value returned by the function DATE_DIFF_STR, which computes the number of days between the order date and the ship date. The days value is then referenced multiple times in the remainder of the query. By giving a name to the value of the expression, the LET clause allows the expression to be written into the query only once. This makes the query shorter, easier to understand, and less error-prone.

Note that there are some orders in the dataset that have an order date but not a ship date. For these orders, the value of days is missing, the days > 2 test in the WHERE clause also returns missing, and the order does not appear in the query result.

WHERE Clause

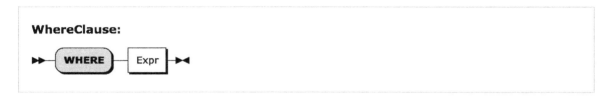

The WHERE clause is familiar to every SQL user, and we've already seen several examples of how it is used. The purpose of a WHERE clause is to operate on the stream of binding tuples generated by the FROM clause, filtering out the tuples that do not satisfy a certain condition. The condition is specified by an expression based on the variable names in the binding tuples. If the expression evaluates to true, the tuple remains in the stream; if it evaluates to anything else, including null or missing, it is filtered out. The surviving tuples are then passed along to the next clause to be processed (usually either GROUP BY or SELECT).

Often, the expression in a WHERE clause is some kind of comparison like quantity > 100. However, any kind of expression is allowed in a WHERE clause. The only thing that matters is whether the expression returns true or not.

Grouping

Grouping is especially important when manipulating hierarchies like the ones that are often found in JSON data. Often you will want to generate output data that includes both summary data and line items within the summaries. For this purpose, N1QL supports several important extensions to the traditional grouping features of SQL. The familiar GROUP BY and HAVING clauses are still there, and they are joined by a new clause called GROUP AS. We'll illustrate these clauses by a series of examples.

GROUP BY Clause

GroupByClause:

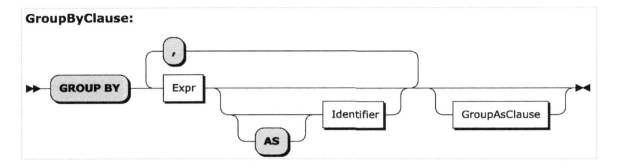

We'll begin our discussion of grouping with an example that you'll find familiar.

(Q23) List the number of orders placed by each customer who has placed an order.

```
FROM orders AS o
GROUP BY o.custid
SELECT o.custid, COUNT(o.orderno) AS `order count`
ORDER BY o.custid;
```

Result:

```
[ { "custid": "C13",
    "order count": 3
  },
  { "custid": "C31",
    "order count": 1
  },
  { "custid": "C35",
    "order count": 1
  },
  { "custid": "C37",
    "order count": 1
  },
  { "custid": "C41",
    "order count": 2
  }
]
```

The input to a GROUP BY clause is the stream of binding tuples generated by the FROM and WHERE clauses. In this query, before grouping, the variable o is bound to each object in the orders collection in turn.

N1QL evaluates the expression in the GROUP BY clause, called the *grouping expression*, once for each of the binding tuples. It then organizes the results into groups in which the grouping expression has a common value (as defined by the = operator). In this example, the grouping expression is o.custid, and each of the resulting groups is a set of orders that have the same customer id. If necessary, a group is formed for orders in which custid is null, and another group is formed for orders that have no custid. This query uses the aggregating function COUNT(o.orderno), which counts how many order numbers are in each group. If we are sure that each order object has a distinct orderno, we could also simply count the order objects in each group by using COUNT(*) in place of COUNT(o.orderno).

In the GROUP BY clause, you may optionally define an alias for the grouping expression. For example, in Q23, you could have written GROUP BY o.custid AS cid. The alias cid could then be used in place of the grouping expression in later clauses. In cases where the grouping expression contains an operator, it is especially helpful to define an alias (for example, GROUP BY salary + bonus AS pay).

Q23 had a single grouping expression, o.custid. If a query has multiple grouping expressions, the combination of grouping expressions is evaluated for every binding tuple, and the stream of binding tuples is partitioned into groups that have values in common for all of the grouping expressions. We'll see an example of such a query in Q24.

After grouping, the number of binding tuples is reduced: instead of a binding tuple for each of the input objects, there is a binding tuple for each group. The grouping expressions (identified by their aliases, if any) are bound to the results of their evaluations. However, all the *non-grouping fields* (that is, fields that were not named in the grouping expressions), are accessible only in a special way: as an argument of one of the *special aggregation functions*: SUM, AVG, MAX, MIN, and COUNT. The clauses that come after grouping can access only properties of groups, including the grouping expressions and aggregate properties of the groups such as COUNT(o.orderno) or COUNT(*). (We'll see an exception when we discuss the new GROUP AS clause.)

You may notice that the results of Q23 do not include customers who have no orders. If we want to include these customers, we need to use an outer join between the `customers` and `orders` collections. This is illustrated by the following example, which also includes the name of each customer.

(Q24) List the customer id and name of each customer, and the number of orders placed by each customer.

```
FROM customers AS c
      LEFT OUTER JOIN orders AS o ON c.custid = o.custid
GROUP BY c.custid, c.name
SELECT c.custid, c.name, COUNT(o.orderno) AS `order count`
ORDER BY c.custid;
```

Result:

```
[ { "custid": "C13",
    "name": "T. Cruise",
    "order count": 3
  },
  { "custid": "C25",
    "name": "M. Streep",
    "order count": 0
  },
  { "custid": "C31",
    "name": "B. Pitt",
    "order count": 1
  },
  ...
]
```

Notice in Q24 what happens when the special aggregation function COUNT is applied to a collection that does not exist, such as the orders of M. Streep: it returns zero. This behavior is unlike that of the other special aggregation functions SUM, AVG, MAX, and MIN, which return `null` if their operand does not exist. This should make you cautious about the COUNT function: If it returns zero, that may mean that the collection you are counting has zero members, or that it does not exist, or that you have misspelled its name.

Q24 also shows how a query block can have more than one grouping expression. In general, the GROUP BY clause produces a binding tuple for each different combination of values for the grouping expressions. In Q24, the `c.custid` field uniquely identifies a customer, so adding `c.name` as a grouping expression does not result in any more groups. Nevertheless, `c.name` must be included as a grouping expression if it is to be referenced outside the GROUP BY clause. If `c.name` were not included in the GROUP BY clause, it would not be a group property and could not be used in the SELECT clause.

Of course, a grouping expression need not be a simple field-name. In Q25, orders are grouped by month using a temporal function to extract the month component of the order dates. In cases like this, it is helpful to define an alias for the grouping expression so it can be referenced elsewhere in the query.

(Q25) List the number of orders in each month in 2017.

```
FROM orders AS o
WHERE DATE_PART_STR(o.order_date, "year") = 2017
GROUP BY DATE_PART_STR(o.order_date, "month") AS month
SELECT month, COUNT(*) AS order_count
ORDER BY month;
```

Result:

```
[ { "month": 4,
    "order_count": 1
  },
  { "month": 5,
    "order_count": 1
  },
  . . .
]
```

In the previous examples, groups were formed from named collections like `customers` and `orders`. But in some queries you need to form groups from a collection that is nested inside another collection, such as `items` inside `orders`. In N1QL, you can do this by using left-correlation in the FROM clause to unnest the inner collection, joining the inner collection with the outer collection, and then performing the grouping on the join, as illustrated in Q26.

(Q26) List the total revenue (sum of quantity times price of items) for each order.

```
FROM orders AS o, o.items as i
GROUP BY o.orderno
SELECT o.orderno, SUM(i.qty * i.price) AS revenue
ORDER BY o.orderno;
```

Result:

```
[ { "orderno": 1001,
    "revenue": 157.73
  },
  { "orderno": 1002,
    "revenue": 10906.55
  },
  { "orderno": 1003,
    "revenue": 477.95
  },
  ...
]
```

After a GROUP BY clause, you have another opportunity to use a LET clause to define some additional variables. A LET clause that follows GROUP BY looks just like a LET clause that follows FROM, and has the same purpose. For example, let's suppose that, in Q26, you want to list only those orders that have revenue greater than 1000, and you want to list them in descending order of revenue. Revenue is defined by the expression SUM(i.qty * i.price), but you probably don't want to type this expression three times, and you certainly don't want the system to execute it three times on the same input. You can solve this problem by using a LET clause to define revenue, as follows:

(Q27) List the total revenue (sum of quantity times price of items) for each order whose revenue is greater than 1000.

```
FROM orders AS o, o.items as i
GROUP BY o.orderno
LET revenue = SUM(i.qty * i.price)
HAVING revenue > 1000
SELECT o.orderno, revenue
ORDER BY revenue DESC;
```

Result:

```
[ { "orderno": 1006,
    "revenue": 18847.58
  },
  { "orderno": 1002,
    "revenue": 10906.55
  },
  { "orderno": 1005,
    "revenue": 4639.92
  },
  { "orderno": 1008,
    "revenue": 1999.80
  }
]
```

You'll often find that join-and-group queries like Q26 and Q27 can be written alternatively as a query block with a subquery, as in the following example. Following what we have learned about subqueries, we use SELECT VALUE inside the subquery and [0] to extract a value from the collection returned by the subquery. Note that the condition in the HAVING clause of Q27 has moved to the WHERE clause of Q28.

(Q28) Alternative form of Q27 (same result).

```
FROM orders AS o
LET revenue =
    (FROM o.items AS i
    SELECT VALUE SUM(i.qty * i.price))[0]
WHERE revenue > 1000
```

```
SELECT o.orderno, revenue
ORDER BY revenue DESC;
```

To review what we've learned, we'll write a query that combines a three-way join with unnesting an array. This query has four grouping expressions, three of which are included simply because they are needed in later clauses. It also defines some aliases that will appear as field-names in the query result: `order no.` is an alias for o.orderno, and `amount due` is an alias for SUM(i.qty * i.price). The back-ticks on the aliases are necessary because they contain blank spaces and punctuation.

(Q29) Create an invoice for each order, showing the order number, date, customer name, address, and total amount due.

```
FROM orders AS o, o.items AS i, customers AS c
WHERE o.custid = c.custid
GROUP BY o.orderno AS `order no.`,
         o.order_date AS date, c.name, c.address
SELECT `order no.`, date, c.name, c.address,
       SUM(i.qty * i.price) AS `amount due`
ORDER BY `order no.`;
```

Result:

```
[ { "order no.": 1001,
    "date": "2017-04-29",
    "name": "R. Duvall",
    "address":
       { "city": "St. Louis, MO",
          "street": "150 Market St.",
          "zipcode": "63101"
       },
    "amount due": 157.73
  },
  ...
]
```

HAVING Clause

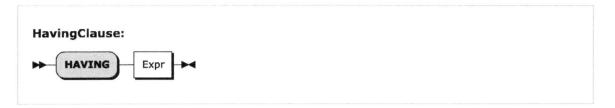

HavingClause:

As you know, the HAVING clause is very similar to the WHERE clause, except that it comes after GROUP BY and applies a filter to groups rather than to individual objects. Here's an example of a HAVING clause that filters orders by applying a condition to their nested arrays of items.

(Q30) Find the order number, date, and revenue of all orders whose revenue (sum over items of quantity times price) is greater than 10000, in descending order by revenue.

```
FROM orders AS o, o.items AS i
GROUP BY o.orderno, o.order_date
HAVING SUM(i.qty * i.price) > 10000
SELECT o.orderno AS order_number, o.order_date,
       SUM(i.qty * i.price) AS revenue
ORDER BY revenue DESC;
```

Result:

```
[ { "order_number": 1006,
    "order_date": "2017-09-02",
    "revenue": 18847.58   },
  { "order_number": 1002,
    "order_date": "2017-05-01",
    "revenue": 10906.55
  }
]
```

Notice that, in Q30, the expression SUM(`i.qty * i.price`) is repeated twice. It would be more elegant to use a LET clause to define an alias for this expression and then use the alias wherever the expression is needed.

(Q31) Alternate form of Q30, using a LET clause to avoid duplicate expressions (same result).

```
FROM orders AS o, o.items AS i
GROUP BY o.orderno, o.order_date
LET revenue =  SUM(i.qty * i.price)
HAVING revenue > 10000
SELECT o.orderno AS order_number, o.order_date, revenue
ORDER BY revenue DESC;
```

Also notice that, if you like the word UNNEST, you could have used it in either of the last two queries without changing their result. Simply change the FROM clause of each query to FROM orders AS o UNNEST o.items AS i.

Aggregation Pseudo-Functions

SQL provides five special functions for performing aggregations on groups: SUM, AVG, MAX, MIN, and COUNT (some implementations provide more). These same functions are supported in N1QL. However, it's worth spending some time on these special functions because they don't behave like ordinary functions. In fact, I'm going to call them "pseudo-functions" because they don't evaluate their operands in the same way as ordinary functions. To see the difference, consider these two examples, which are syntactically similar:

Example 1: `SELECT LENGTH(name) FROM customers`

In Example 1, LENGTH is an ordinary function. It simply evaluates its operand (name) and then returns a result computed from the operand.

Example 2: `SELECT AVG(rating) FROM customers`

The effect of AVG in Example 2 is quite different. Rather than performing a computation on an individual `rating` value, AVG has a global effect: it effectively restructures the query. As a pseudo-function, AVG requires its operand to be a group; therefore, it automatically collects all the `rating` values from the query block and forms them into a group.

The aggregation pseudo-functions always require their operand to be a group. In some queries, the group is explicitly generated by a GROUP BY clause, as in Q32:

(Q32) List the average credit rating of customers by zipcode.

```
FROM customers AS c
GROUP BY c.address.zipcode AS zip
SELECT zip, AVG(c.rating) AS `avg credit rating`
ORDER BY zip;
```

Result:

```
[ { "avg credit rating": 625
  },
  { "zip": "02115",
    "avg credit rating": 657.5
  },
  { "zip": "02340",
    "avg credit rating": 690
  },
  { "zip": "63101",
    "avg credit rating": 695
  }
]
```

Note in the result of Q32 that one or more customers had no zipcode. These customers were formed into a group for which the value of the grouping key is missing. When the query results were returned in JSON format, the missing key simply does not appear. Also note that the group whose key is missing appears first because missing is considered to be smaller than any other value. If some customers had had null as a zipcode, they would have been included in another group, appearing after the missing group but before the other groups.

When an aggregation pseudo-function is used without an explicit GROUP BY clause, it implicitly forms the entire query block into a single group, as in Q33:

(Q33) Find the average credit rating among all customers.

```
FROM customers AS c
SELECT AVG(c.rating) AS `avg credit rating`;
```

Result:

```
[ { "avg credit rating": 670 } ]
```

The aggregation pseudo-function COUNT has a special form in which its operand is * instead of an expression. For example, SELECT COUNT(*) FROM customers simply returns the total number of customers, whereas SELECT COUNT(rating) FROM customers returns the number of customers who have known ratings (that is, their ratings are not null or missing).

Because the aggregation pseudo-functions sometimes restructure their operands, they can be used only in query blocks where (explicit or implicit) grouping is being done. Therefore the pseudo-functions cannot operate directly on arrays or multisets. For operating directly on JSON collections, N1QL provides a set of ordinary aggregation functions. Each ordinary aggregation function (except the one corresponding to COUNT) has two versions: one that ignores null and missing values and one that returns null if a null or missing value is encountered anywhere in the collection. The names of the aggregation functions are as follows:

Aggregation pseudo-function; operates on groups only	Ordinary function; operates on any collection; ignores nulls	Ordinary function; operates on any collection; returns null if null is encountered
SUM	ARRAY_SUM	STRICT_SUM
AVG	ARRAY_AVG	STRICT_AVG
MAX	ARRAY_MAX	STRICT_MAX
MIN	ARRAY_MIN	STRICT_MIN
COUNT	ARRAY_COUNT	STRICT_COUNT (see exception below)

Exception: the ordinary aggregation function STRICT_COUNT operates on any collection, and returns a count of its items, including null values in the count. In this respect, STRICT_COUNT is more similar to COUNT(*) than to COUNT(expression).

Note that the ordinary aggregation functions that ignore nulls have names beginning with "ARRAY." This naming convention has historical roots. Despite their names, the functions operate on both arrays and multisets.

Because of the special properties of the aggregation pseudo-functions, SQL (and therefore N1QL) is not a pure functional language. But every query that uses a pseudo-function can be rewritten into an equivalent query that uses an ordinary function. In fact, that is how the pseudo-functions are implemented in N1QL: queries that use them are translated into pure functional versions before execution. Q34 and Q35 are examples of how queries can be rewritten to avoid the use of pseudo-functions.

(Q34) Alternative form of Q33, using the ordinary function ARRAY_AVG rather than the aggregating pseudo-function AVG.

```
SELECT ARRAY_AVG(
    (SELECT VALUE c.rating
     FROM customers AS c) ) AS `avg credit rating`;
```

Result (same as Q33):

```
[ { "avg credit rating": 670 } ]
```

If the function STRICT_AVG had been used in Q34 in place of ARRAY_AVG, the average credit rating returned by the query would have been null, because at least one customer has no credit rating.

(Q35) Alternative form of Q30, using the ordinary function ARRAY_SUM rather than the aggregating pseudo-function SUM.

```
FROM orders AS o
LET revenue = ARRAY_SUM(
        (FROM o.items AS i
         SELECT VALUE i.qty * i.price) )
WHERE revenue > 10000
SELECT o.orderno AS order_number, o.order_date, revenue
ORDER BY revenue DESC;
```

Result (same as Q30):

```
[ { "order_number": 1006,
    "order_date": "2017-09-02",
    "revenue": 18847.58
  },
  { "order_number": 1002,
    "order_date": "2017-05-01",
    "revenue": 10906.55
  }
]
```

Both Q34 and Q35 have a subquery inside an ordinary aggregation function. Note that double parentheses are used in each case: one set of parentheses to enclose the argument of the function, and another pair to enclose the subquery. You need both pairs whenever the argument of a function happens to be a subquery.

Here's another example in which an ordinary aggregation function operates directly on an existing collection (named items) rather than on the result of a subquery:

(Q36) List all the orders received in September 2017, with the number of line items in each order.

```
FROM orders AS o
WHERE DATE_PART_STR(o.order_date, "year") = 2017
AND DATE_PART_STR(o.order_date, "month") = 9
SELECT o.orderno, ARRAY_COUNT(o.items) AS line_items
ORDER BY o.orderno;
```

Result:

```
[
  { "orderno": 1006,
    "line_items": 3
  },
  { "orderno": 1007,
```

```
        "line_items": 2
    }
]
```

Query Q36 would have been more difficult to write (and probably more expensive to execute) using the SQL pseudo-function COUNT. Here's what it would look like:

(Q37) Alternate form of Q36, using COUNT rather than ARRAY_COUNT.

```
FROM orders AS o, o.items AS i
WHERE DATE_PART_STR(o.order_date, "year") = 2017
AND DATE_PART_STR(o.order_date, "month") = 9
GROUP BY o.orderno
SELECT o.orderno, COUNT(i) AS line_items
ORDER BY o.orderno;
```

Result: Same as Q36

GROUP AS Clause

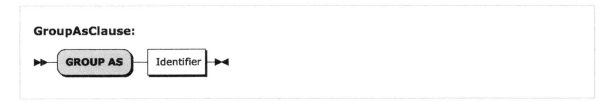

GroupAsClause:

All the query results we have seen so far have been flat sequences of objects. We have not produced any hierarchical results (except for hierarchies that were already present in the input data). But JSON is a hierarchical format, and a fully featured JSON query language needs to be able to produce hierarchies of its own, with computed data at every level of the hierarchy. The key feature of N1QL that makes this possible is the GROUP AS clause.

A query may have a GROUP AS clause only if it has a GROUP BY clause. As we have seen, the GROUP BY clause "hides" the original objects in each group, exposing only the grouping expressions and special aggregation functions on the non-grouping fields. The purpose of the GROUP AS clause is to make the original objects in the group visible to subsequent clauses. Thus the query can generate output data both for the group as a whole and for the individual objects inside the group.

For each group, the GROUP AS clause preserves all the objects in the group, just as they were before grouping, and gives a name to this preserved group. The group name can then be used in the FROM clause of a subquery to process and return the individual objects in the group.

To see how this works, we'll write some queries that investigate the customers in each zipcode and their credit ratings. This would be a good time to review the sample database in Appendix A. I'll briefly summarize some of the data here:

> Customers in zipcode 02115:
> > C35, J. Roberts, rating 565
> > C37, T. Hanks, rating 750
>
> Customers in zipcode 02340:
> > C25, M. Streep, rating 690
>
> Customers in zipcode 63101:
> > C13, T. Cruise, rating 750
> > C31, B. Pitt, (no rating)
> > C41, R. Duvall, rating 640
>
> Customers with no zipcode:
> > C47, S. Loren, rating 625

Now let's consider the effect of the following clauses:

```
FROM customers AS c
GROUP BY c.address.zipcode
GROUP AS g
```

This query fragment iterates over the customers objects, using the iteration variable c. The GROUP BY clause forms the objects into groups, each with a common zipcode (including one group for customers with no zipcode). After the GROUP BY clause, we can see the grouping expression, c.address.zipcode, but other fields such as c.custid and c.name are visible only to special aggregation functions.

The clause GROUP AS g now makes the original objects visible again. For each group in turn, the variable g is bound to a multiset of objects, each of which has a field named c, which in turn contains one of the original objects. Thus after GROUP AS g, for the group with zipcode 02115, g is bound to the following multiset:

```
[ { "c": { "custid": "C35",
            "name": "J. Roberts",
            "address":
                { "street": "420 Green St.",
                  "city": "Boston, MA",
                  "zipcode": "02115"
                },
            "rating": 565
          }
  },
  { "c": { "custid": "C37",
            "name": "T. Hanks",
            "address":
                { "street": "120 Harbor Blvd.",
                  "city": "St. Louis, MO",
                  "zipcode": "02115"
                },
            "rating": 750
          }
  }
]
```

Thus, the clauses following GROUP AS can see the original objects by writing subqueries that iterate over the multiset g.

Now you are wondering why the extra level named c was introduced into this multiset. Why not let the name g directly represent the multiset of original objects? The reason is because the groups might have been formed from a join of two or more collections. Suppose that the FROM clause looked like FROM customers AS c, orders AS o. Then each item in the group would contain both a `customers` object and an `orders` object, and these two objects might both have a field with the same name. To avoid ambiguity, each of the original objects is wrapped in an "outer" object that gives it the name of its iteration variable in the FROM clause. Consider this fragment:

```
FROM customers AS c, orders AS o
WHERE c.custid = o.custid
GROUP BY c.address.zipcode
GROUP AS g
```

In this case, following GROUP AS g, the variable g would be bound to the following collection:

```
[ { "c": { an original customers object },
    "o": { an original orders object }
  },
  { "c": { another customers object },
    "o": { another orders object }
  },
  . . .
]
```

After using GROUP AS to make the content of a group accessible, you will probably want to write a subquery to access that content. A subquery for this purpose is written in exactly the same way as any other subquery. The name you specified in the GROUP AS clause (g in the above example) is the name of a collection of objects. You can write a FROM clause to iterate over the objects in the collection, and you can specify an iteration variable to represent each object in turn. For GROUP AS queries in this tutorial, I'll use g as the name of the reconstituted group, and gi as an iteration variable representing one object inside the group. Of course, you can use any names you like for these purposes.

Now we are ready to take a look at how GROUP AS might be used in a query. Suppose that we want to group customers by zipcode, and for each group we want to see the average credit rating and a list of the individual customers in the group. Here's a query that does that:

(Q38a) For each zipcode, list the average credit rating in that zipcode, followed by the customer numbers and names in numeric order.

```
FROM customers AS c
GROUP BY c.address.zipcode AS zip
GROUP AS g
SELECT zip, AVG(c.rating) AS `avg credit rating`,
    (FROM g AS gi
     SELECT gi.c.custid, gi.c.name
     ORDER BY gi.c.custid) AS `local customers`
ORDER BY zip;
```

Result:

```
[ { "avg credit rating": 625,
    "local customers": [
       { "custid": "C47",
         "name": "S. Loren"
       }
    ]
  },
  { "zip": "02115",
    "avg credit rating": 657.5,
    "local customers": [
       { "custid": "C35",
         "name": "J. Roberts"
       },
       { "custid": "C37",
         "name": "T. Hanks"
       }
    ]
  },
  { "zip": "02340",
    "avg credit rating": 690,
    "local customers": [
       { "custid": "C25",
         "name": "M. Streep"
       }
```

```
            ]
        },
        { "zip": "63101",
          "avg credit rating": 695,
          "local customers": [
              { "custid": "C13",
                "name": "T. Cruise"
              },
              { "custid": "C31",
                "name": "B. Pitt"
              },
              { "custid": "C41",
                "name": "R. Duvall"
              }
          ]
        }
    ]
```

Note that this query contains two ORDER BY clauses: one in the outer query and one in the subquery. These two clauses govern the ordering of the outer-level list of zipcodes and the inner-level lists of customers, respectively. Also note that the group of customers with no zipcode comes first in the output list.

In Q38a, I used the explicit iteration variable `gi` in the FROM clause of the subquery over groups. That's consistent with my general advice about writing FROM clauses. Of course, you could have omitted the explicit variable, in which case the FROM clause of the subquery would have defined an implicit variable named g, and field references like `c.custid` would have implicitly represented fields inside the object bound to g. Q38b shows what the query would have looked like in this case. The tradeoff is up to you: brevity (Q38b) vs. explicit detail (Q38a). In the remainder of this tutorial, I'll use explicit iteration variables in the FROM clauses of both subqueries and outer query blocks to clarify how the queries are actually executed.

(Q38b) Alternative version of Q38a, omitting explicit iteration variables in the subquery.

```
FROM customers AS c
GROUP BY c.address.zipcode AS zip
```

```
GROUP AS g
SELECT zip, AVG(c.rating) AS `avg credit rating`,
    (FROM g
     SELECT c.custid, c.name
     ORDER BY c.custid) AS `local customers`
ORDER BY zip;
```

Result: Same as Q38a

We can make this example more interesting by using a correlated subquery. As an SQL user, you know that a correlated subquery is a subquery that contains a reference to a value defined in the outer query. In Q39, the outer query defines a value named `best rating`, and this value is referenced by the subquery.

> (Q39) For each zipcode, list the highest credit rating in that zipcode, followed by the customer ids and names of the customers who have that rating.

```
FROM customers AS c
GROUP BY c.address.zipcode AS zip
GROUP AS g
LET `best rating` = MAX(c.rating)
SELECT zip, `best rating`,
    (FROM g AS gi
     WHERE gi.c.rating = `best rating`
     SELECT gi.c.custid, gi.c.name
     ORDER BY gi.c.custid) AS `best customers`
ORDER BY zip;
```

In reading the subquery, remember that g represents a group of objects; gi is an iteration variable bound to one of those objects; and gi.c.rating represents the value c.rating inside the gi object.

Result:

```
[
  { "best rating": 625,
```

```
        "best customers": [
          { "custid": "C47",
            "name": "S. Loren"
          }
        ]
      },
      { "zip": "02115",
        "best rating": 750,
        "best customers": [
          { "custid": "C37",
            "name": "T. Hanks"
          }
        ]
      },
      { "zip": "02340",
        "best rating": 690,
        "best customers": [
          { "custid": "C25",
            "name": "M. Streep"
          }
        ]
      },
      { "zip": "63101",
        "best rating": 750,
        "best customers": [
          { "custid": "C13",
            "name": "T. Cruise"
          }
        ]
      }
    ]
```

Note that Q39 has created a new hierarchy in which zip is the root. The outer query block groups the customers by zipcode and finds the highest rating in each zipcode; then the correlated subquery scans each group to find all the customers whose rating matches the highest rating.

In Q39, the correlated subquery is in the SELECT clause of the outer query block. Q39 might have been expressed more elegantly by moving the correlated subquery to the LET clause and binding an identifier to the result of the subquery, as in Q40:

(Q40) Alternate form of Q39, moving subquery to LET clause.

```
FROM customers AS c
GROUP BY c.address.zipcode AS zip
GROUP AS g
LET `best rating` = MAX(c.rating),
    `best customers` =
        (FROM g AS gi
          WHERE gi.c.rating = `best rating`
          SELECT gi.c.custid, gi.c.name
          ORDER BY gi.c.custid)
SELECT zip, `best rating`, `best customers`
ORDER BY zip;
```

Result: Same as Q39

As noted in an earlier section, any query containing a special aggregation function can be rewritten to a query that uses no special functions. A GROUP AS clause is often helpful in this process, as illustrated by Q41, which rewrites Q32 to eliminate the special AVG function.

(Q32, repeated for convenience) Find the average credit rating of customers by zipcode.

```
FROM customers AS c
GROUP BY c.address.zipcode AS zip
SELECT zip, AVG(c.rating) AS `avg credit rating`
ORDER BY zip;
```

(Q41) Alternate form of Q32, avoiding use of special AVG function.

```
FROM customers AS c
GROUP BY c.address.zipcode AS zip
GROUP AS g
```

```
SELECT zip,
    ARRAY_AVG((FROM g AS gi SELECT VALUE gi.c.rating))
        AS `avg credit rating`
ORDER BY zip;
```

Note the double parentheses following ARRAY_AVG: one pair to enclose the argument of the function, and another pair to enclose the subquery.

Result: Same as Q32 (repeated here for convenience)

```
[ { "avg credit rating": 625
  },
  { "zip": "02115",
    "avg credit rating": 657.5
  },
  { "zip": "02340",
    "avg credit rating": 690
  },
  { "zip": "63101",
    "avg credit rating": 695
  }
]
```

Our next example will show how to use GROUP BY and GROUP AS to filter an existing hierarchy.

(Q42) For each day in which the total revenue (sum of quantity times price for all items ordered on that day) is greater than 1000.00, list the date, total revenue, and all the ordered items whose price is more than 100.00.

```
FROM orders AS o, o.items AS i
GROUP BY o.order_date
GROUP AS g
LET revenue = SUM(i.price * i.qty)
HAVING revenue > 1000.00
SELECT o.order_date AS `good day`, revenue,
    (FROM g AS gi
```

```
      WHERE gi.i.price > 100.00
      SELECT gi.i.itemno, gi.i.price
      ORDER BY gi.i.itemno) AS `expensive items`
ORDER BY o.order_date;
```

Result:

```
[ { "good day": "2017-05-01",
    "revenue": 10906.55,
    "expensive items": [
      {
        "itemno": 460,
        "price": 100.99
      }
    ]
  },
  { "good day": "2017-08-30",
    "revenue": 4639.92,
    "expensive items": [
      {
        "itemno": 375,
        "price": 149.98
      },
      {
        "itemno": 780,
        "price": 1500
      }
    ]
  },
  { "good day": "2017-09-02",
    "revenue": 18847.58,
    "expensive items": []
  },
  { "good day": "2017-10-13",
    "revenue": 1999.80,
    "expensive items": []
  }
]
```

Note that we have aggregated the orders by `order_date`, and then filtered the resulting hierarchy at two levels, retaining only dates on which revenue was greater than 1000.00, and only those items with price greater than 100.00. We have also added a new computed field, `revenue`, to the query result. Note that on two of the "good days" (revenue > 1000) there were no "expensive items" (price > 100).

In our next example, we'll combine a three-way join with GROUP AS. This query will make it clear why, in the multiset generated by GROUP AS, each part of the join is wrapped in its own object.

(Q43) For each customer who has placed an order, show the customer id and name, followed by all the items ordered by that customer, showing the date, item number, and quantity for each. Order the customers by customer id, and the items ordered by each customer by date and item number.

```
FROM customers AS c, orders AS o, o.items AS i
WHERE c.custid = o.custid
GROUP BY c.custid, c.name
GROUP AS g
SELECT c.custid, c.name,
    (FROM g AS gi
     SELECT gi.o.order_date, gi.i.itemno, gi.i.qty
     ORDER BY gi.o.order_date, gi.i.itemno) AS recent_items
ORDER BY c.custid;
```

Result:

```
[ { "custid": "C13",
    "name": "T. Cruise",
    "recent_items":
       [ { "order_date": "2017-05-01",
           "itemno": 460,
           "qty": 95
         },
         { "order_date": "2017-05-01",
           "itemno": 680,
```

```
            "qty": 150
        },
        { "order_date": "2017-09-13",
          "itemno": 185,
          "qty": 5
        },
        { "order_date": "2017-09-13",
          "itemno": 680,
          "qty": 1
        },
        { "order_date": "2017-10-13",
          "itemno": 460,
          "qty": 20
        }
      ]
    },
    ...
]
```

Notice that, in the subquery, we referenced fields from `orders` (`gi.o.order_date`) and fields from `items` (`gi.i.itemno`, `gi.i.qty`). The aliases o and i in these path expressions resolve any ambiguities that might arise from duplicate field names.

The result of Q43 did not include any customers who had no orders. If we want to include these customers, we can revise the query to use a left outer join. This example also illustrates the use of a comment inside a query. Note that the customer with id C25 is included in the result of Q44 but not in Q43.

(Q44) Alternate form of Q43, using left outer join to include all customers.

```
FROM customers AS c LEFT OUTER JOIN
    /* This subquery unnests the items in each order,
         returning an array of results named sq */
    (FROM orders AS o, o.items AS i
      SELECT o.custid, o.order_date, i.itemno, i.qty) AS sq
ON c.custid = sq.custid
GROUP BY c.custid, c.name
```

```
        GROUP AS g
        SELECT c.custid, c.name,
              (FROM g AS gi
               SELECT gi.sq.order_date, gi.sq.itemno, gi.sq.qty
               ORDER BY gi.sq.order_date, gi.sq.itemno)
               AS recent_items
        ORDER BY c.custid;
```

Result:

```
[ { "custid": "C13",
    "name": "T. Cruise",
    "recent_items":
        [ { "order_date": "2017-05-01",
            "itemno": 460,
            "qty": 95
          },
          { "order_date": "2017-05-01",
            "itemno": 680,
            "qty": 150
          },
          { "order_date": "2017-09-13",
            "itemno": 185,
            "qty": 5
          },
          { "order_date": "2017-09-13",
            "itemno": 680,
            "qty": 1
          },
          { "order_date": "2017-10-13",
            "itemno": 460,
            "qty": 20
          }
        ]
  },
  { "custid": "C25",
    "name": "M. Streep",
```

```
    "recent_items": [ { } ]
  },
  ...
]
```

The next example illustrates how GROUP AS might be used to invert a hierarchy. In the `orders` dataset, `custid` is a field at the top of the hierarchy and `itemno` is a field of a nested object. This example inverts the hierarchy to put `itemno` at the top level and `custid` at a nested level.

(Q45) For each item number, list the total quantity of that item that has been ordered, with a list of the customer ids who ordered that item and the dates of the orders, in chronological order.

```
FROM orders AS o, o.items AS i
GROUP BY i.itemno
GROUP AS g
LET total_on_order = SUM(i.qty)
SELECT i.itemno, total_on_order,
    (FROM g AS gi
     SELECT gi.o.order_date, gi.o.custid
     ORDER BY gi.o.order_date) AS purchasers
ORDER BY i.itemno;
```

Result:

```
[ { "itemno": 120,
    "total_on_order": 67,
    "purchasers": [
      { "custid": "C31",
        "order_date": "2017-06-15"
      },
      { "custid": "C41",
        "order_date": "2017-09-02"
      }
    ]
  },
  { "itemno": 185,
```

```
      "total_on_order": 5,
      "purchasers": [
        { "custid": "C13",
          "order_date": "2017-09-13"
        }
      ]
    },
  ...
  ]
```

If you've gotten this far, you've learned a lot about manipulating hierarchies with SQL++. You've filtered hierarchies, inverted hierarchies, and created new hierarchies. You've come a long way from querying flat tables!

SELECT Clause

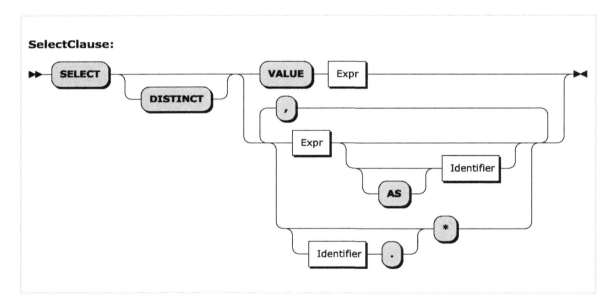

You may think of a SELECT clause as the first line in an SQL query block, but there's a reason why I've waited until now to discuss it. The clauses we've discussed so far really do nothing but bind some variables to values, setting up an environment for the SELECT clause. The job of the SELECT clause is to actually generate the result of the query block, using the variable bindings that have been created by other clauses.

As we've seen in an earlier section, there are two kinds of SELECT clauses: the "ordinary" one you are familiar with as an SQL user, and a new kind that begins with SELECT VALUE.

To summarize, a "ordinary" SELECT clause consists of a comma-separated list of expressions, with an optional alias for each expression. The SELECT clause is invoked once for each set of bindings generated by the FROM (and maybe GROUP BY/GROUP AS) clauses. For each set of bindings, the SELECT clause generates a JSON object. The names of the object fields are the aliases of the SELECT expressions, and their values are the values of the SELECT expressions. No ordering is defined among the fields of the objects. The result of the query block is a multiset (or array, if ORDER BY is specified) containing these objects. When the query result is serialized for output in JSON format, multisets are serialized as arrays.

The newer SELECT VALUE clause may stand alone or it may be part of a query block with a FROM clause. In either case, the SELECT VALUE clause contains a single expression. If the SELECT VALUE stands alone, it simply evaluates the expression and returns a multiset containing the value of the expression. If it is part of a query block, the SELECT VALUE clause evaluates the expression for each set of bindings generated by the other clauses, and returns the results as a multiset (or array, if ORDER BY is specified). SELECT VALUE is especially useful in subqueries where you want the subquery to return a simple value that is not wrapped in an object.

In place of an expression, the writer of a SELECT clause may use the symbol *, which means "all the variables that are currently defined." The symbol * preceded by an identifier, as in c . *, means "all the fields of the object bound to variable c." We'll describe SELECT * queries more fully in a later section.

As in SQL, a SELECT clause (with or without VALUE) may contain the optional keyword DISTINCT. There are no surprises here: the DISTINCT keyword causes duplicates to be eliminated from the result of the query block. The following query illustrates the use of DISTINCT in a SELECT VALUE query. Although T. Cruise has multiple orders for some item numbers, each item number appears only once in the query result.

(Q46) Make a list of all the different items ordered by T. Cruise.

```
SELECT DISTINCT VALUE i.itemno
FROM orders AS o, o.items AS i
WHERE o.custid IN
   (SELECT VALUE c.custid
    FROM customers AS c
    WHERE c.name = "T. Cruise");
```

Result: [185, 460, 680]

An ordinary SELECT clause always returns a collection of objects. That doesn't give you a lot of flexibility. A SELECT VALUE clause gives you more power to format your query result as you see fit. You can wrap the result in an object constructor to give it a label. You can also use constructors to build objects that didn't exist before. To illustrate this point, we'll revisit a query we've seen before, and see how it could be rewritten using SELECT VALUE.

> (Q47) (Similar to Q32) List the average credit rating of customers in each zipcode, rounded to the nearest integer.

```
FROM customers AS c
GROUP BY c.address.zipcode AS zip
SELECT zip, ROUND(AVG(c.rating)) AS avg_rating
ORDER BY zip;
```

Result:

```
[ { "avg_rating": 625
  },
  { "zip": "02115",
    "avg_rating": 658
  },
  { "zip": "02340",
    "avg_rating": 690
  },
  { "zip": "63101",
    "avg_rating": 695
  }
]
```

Note that the customers with no zipcode have a group of their own, and that this group appears first in the ordering of zipcodes.

The next example uses SELECT VALUE to compute a similar result with a label and a more compact organization. Note that in the objects returned by this query, both the field-names and the field-values are computed. Note also that we have added an IS KNOWN test to eliminate customers with no zipcode. If we had not done this, the query would have tried to create an object field whose name is an empty string, which is not allowed.

(Q48) List the average credit rating of customers in each valid zipcode, rounded to the nearest integer (alternative format).

```
SELECT VALUE { "Average credit rating by zipcode" :
   (FROM customers AS c
    WHERE c.address.zipcode IS KNOWN
    GROUP BY c.address.zipcode AS zip
    SELECT VALUE { zip : ROUND(AVG(c.rating)) }
    ORDER BY zip
   )
};
```

Result:

```
[ { "Average credit rating by zipcode": [
      { "02115": 658 },
      { "02340": 690 },
      { "63101": 695 }
    ]
  }
]
```

One more note: the expressions in the SELECT clause (and other clauses as well) contain the names of variables that were defined elsewhere in the query. Some of these names may be qualified names like a.b. If you've defined a distinct iteration variable for every collection in your FROM clauses and used these variables to qualify all your field names, the meaning of each name should be quite clear. But the actual rules for interpreting a name can be complex, and awareness of these rules may be helpful in debugging a query that doesn't behave as expected. For your enjoyment, the formal rules for resolving names in N1QL are described in detail in Appendix B.

SELECT *

In SQL, `SELECT *` means "select all the columns." In N1QL, `SELECT *` has a related but slightly different meaning. It means "select all the variables that are currently defined." Let's look at an example to understand how this works.

> (Q49) Show all available information about customers in zipcode 02115.

```
FROM customers AS c
WHERE c.address.zipcode = "02115"
SELECT *;
```

By analogy with SQL, you might expect Q49 to return all the fields from each customer-object that has the desired zipcode. That's almost right. Here's the actual result of Q49:

```
[ { "c": { "custid": "C35",
           "name": "J. Roberts",
           "address": {
                "city": "Boston, MA",
                "street": "420 Green St.",
                "zipcode": "02115"
             },
           "rating": 565
        }
  },
  { "c": { "custid": "C37",
           "name": "T. Hanks",
           "address": {
                "city": "Boston, MA",
                "street": "120 Harbor Blvd.",
                "zipcode": "02115"
             },
           "rating": 750
```

```
            }
        }
    ]
```

During processing of Q49, the SELECT * clause is executed multiple times. For each execution of SELECT *, the variable c is bound to one object from the customers collection that has the desired zipcode. Since c is the only variable that is defined, the SELECT * clause simply returns that variable and the object to which it is bound. That's why, in the result of Q49, you see the variable-name c attached to each of the objects that is returned.

If you would prefer to suppress the variable-name and simply get the fields of the original objects, as SELECT * would have done in SQL, there's a way to do that. Instead of SELECT *, you could write SELECT c.*. The notation c.* means "all the fields of the object bound to variable c" (of course, this works only if c is bound to an object). Here's a version of the query whose result looks more like an SQL result:

(Q50) Repeat Q49, omitting the variable-name from the query result.

```
FROM customers AS c
WHERE c.address.zipcode = "02115"
SELECT c.*;
```

Result:

```
[ { "custid": "C35",
    "name": "J. Roberts",
    "address": {
        "city": "Boston, MA",
        "street": "420 Green St.",
        "zipcode": "02115"
      },
    "rating": 565
  },
  { "custid": "C37",
    "name": "T. Hanks",
    "address": {
```

```
                "city": "Boston, MA",
                "street": "120 Harbor Blvd.",
                "zipcode": "02115"
            },
        "rating": 750
      }
   ]
```

Now you may be wondering about why SELECT * behaves slightly differently in N1QL as compared to SQL, and about the purpose of the variable-dot-star notation. I'll try to deal with these questions by some example queries. The first example query is a join:

> (Q51) Show all available information about customers in zipcode 02115 and their orders.

```
FROM customers AS c, orders AS o
WHERE c.custid = o.custid
AND c.address.zipcode = "02115"
SELECT *;
```

In this query, objects from customers are being joined to objects from orders that have matching custid fields. The result is a collection of pairs. Each pair has two parts: a customers object (bound to c) and an orders object (bound to o). Each of these objects has a custid field. In Q51 the custid fields of the two objects have the same value. But it's possible to join two objects that contain fields with the same name but different values. To preserve the integrity of the two objects and to disambiguate their names, each object is labeled with the name of the variable to which it is bound.

> Result of Q51 (details of ordered items are truncated to save space):

```
[ { "c": { "custid": "C35",
           "name": "J. Roberts",
           "address": {
                "city": "Boston, MA",
                "street": "420 Green St.",
```

```
                    "zipcode": "02115"
                },
                "rating": 565
            },
        "o": { "orderno": 1004,
               "custid": "C35",
               "order_date": "2017-07-10",
               "ship_date": "2017-07-15"
               "items": [ ... ]
            }
    },
    { "c": { "custid": "C37",
             "name": "T. Hanks",
             "address": {
                   "city": "Boston, MA",
                   "street": "120 Harbor Blvd.",
                   "zipcode": "02115"
                },
                "rating": 750
        },
        "o": { "orderno": 1005,
               "custid": "C37",
               "order_date": "2017-08-30",
               "items": [ ... ]
            }
    }
]
```

Next, we'll explore some new uses of the variable-dot-star notation. This notation allows you to "augment" a query result with some fields that were not present in the original objects. Here's a very simple example:

(Q52) For each customer in zipcode 02115, show all fields as well as one additional field, "region": "Northeast".

```
FROM customers AS c
WHERE c.address.zipcode = "02115"
SELECT c.*, "Northeast" AS region;
```

Result:

```
[ { "custid": "C35",
    "name": "J. Roberts",
    "address": {
        "city": "Boston, MA",
        "street": "420 Green St.",
        "zipcode": "02115"
      },
    "rating": 565,
    "region": "Northeast"
  },
  { "custid": "C37",
    "name": "T. Hanks",
    "address": {
        "city": "Boston, MA",
        "street": "120 Harbor Blvd.",
        "zipcode": "02115"
      },
    "rating": 750,
    "region": "Northeast"
  }
]
```

Q52 is not very interesting because it simply adds the same field to every object in the query result. It would be more interesting to augment each object by a different value that was somehow computed from the fields of the object. In the next example, we will augment each output object by an additional computed field called credit.

(Q53) For each customer in zipcode 02115, show all fields as well as an additional credit field computed from other values in the object.

```
FROM customers AS c
LET credit =
    CASE
       WHEN c.rating > 650 THEN "Good"
       WHEN c.rating BETWEEN 500 AND 650 THEN "Fair"
```

```
        ELSE "Poor"
    END
WHERE c.address.zipcode = "02115"
SELECT c.*, credit;
```

Result:

```
[ { "custid": "C35",
    "name": "J. Roberts",
    "address": {
        "city": "Boston, MA",
        "street": "420 Green St.",
        "zipcode": "02115"
      },
    "rating": 565,
    "credit": "Fair"
  },
  { "custid": "C37",
    "name": "T. Hanks",
    "address": {
        "city": "Boston, MA",
        "street": "120 Harbor Blvd.",
        "zipcode": "02115"
      },
    "rating": 750,
    "credit": "Good"
  }
]
```

In Q53, the `credit` field is computed by a simple CASE expression. Of course, in more complex examples the augmented field could be computed by invoking a function, which might depend on fields from multiple objects in the case of a join-query.

Here's a final piece of advice: remember that, in a schemaless environment, you don't have any way of knowing what a SELECT * query will return. Each object returned by such a query may have a different set of fields. That may be okay if you're just exploring a dataset to find out what it contains. But if you're in a production environment that depends on getting a specific result, it's better practice to explicitly name the fields that you expect to be returned.

ORDER BY, LIMIT, and OFFSET

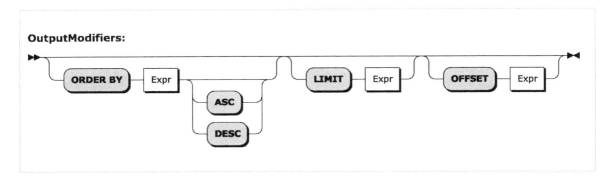

The final three (optional) clauses in a query block are ORDER BY, LIMIT, and OFFSET. As an SQL user, you will not find any surprises here. The ORDER BY clause defines the ordering among objects in the query result, and may trigger a sort. The LIMIT clause places an upper limit on the number of objects to be included in the query result. The OFFSET clause specifies a number of items in the output stream to be discarded before the query result begins.

Since N1QL operates on schemaless data, there is no guarantee that the ordering keys will actually exist in every object in the query result. If an ordering key does not exist in some object, it is considered to have the value missing. For the purpose of ORDER BY, missing is ordered before null, which in turn is ordered before any other value (if DESC is specified, this ordering is reversed). Note that, when an object is serialized into JSON, any field that has the value missing simply disappears. So in your output listing, you may see some objects that do not have any ordering key, and if so, they will appear at the beginning of the listing (or at the end if DESC is specified).

Any query block may have ORDER BY, LIMIT, and/or OFFSET clauses, even if it is in a subquery. In SQL, these clauses were considered to apply to the query as a whole, and were accepted only in the outermost query block. But N1QL is often used to create hierarchical JSON documents in which arrays can be nested inside other arrays. A subquery can be used to generate a nested array that needs to have an ordering of its own.

If the query block has an ORDER BY clause, its result is an array; otherwise its result is a multiset. This distinction is mainly significant for processing of intermediate results during the evaluation of a query. When the final result is delivered in JSON format, a multiset is serialized as an array with arbitrary ordering. If a query block has a LIMIT or OFFSET clause without an ORDER BY clause, the items that are included in the query result are nondeterministic.

Building Queries from Query Blocks

We've finally arrived at the root of the N1QL syntax tree. The following syntax diagram defines a Query. By navigating through all the syntax diagrams, starting with this one, you can see the structure of any query.

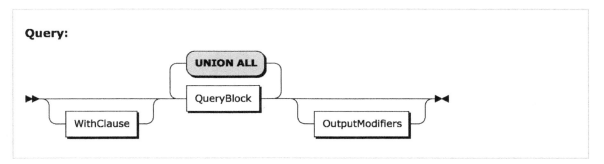

A query block by itself is a valid query. In fact, many queries consist of a single query block. But N1QL also provides some language features that can be used outside a query block to define an environment for the query block or to assemble a query from multiple query blocks.

WITH Clause

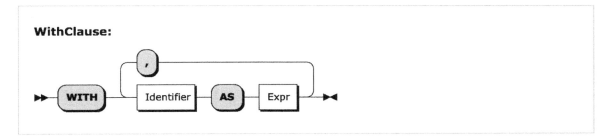

A WITH clause can be used to create a set of variable bindings that are effective throughout a query, even if it consists of multiple query blocks. Here's an example:

```
WITH start_date AS "2017-06-01", end_date AS "2017-12-31"
```

This clause defines two global variables named `start_date` and `end_date` that can be referenced anywhere in the query blocks that follow. You might do this to avoid typing these dates multiple times, to document their meaning, and to make it possible to revise the dates in a single place. You can think of a WITH clause as a global LET clause that applies to an entire query.

A WITH clause often contains a subquery that is needed to compute some result that is used later in the main query. In cases like this, you can think of the WITH clause as computing a "temporary view" of the input data. The next example uses a WITH clause to compute the total revenue of each order in 2017; then the main part of the query finds the minimum, maximum, and average revenue for orders in that year.

(Q54) Find the minimum, maximum, and average revenue among all orders received in the year 2017, rounded to the nearest integer.

```
WITH order_revenue AS
  (FROM orders AS o, o.items AS i
   WHERE DATE_PART_STR(o.order_date, "year") = 2017
   GROUP BY o.orderno
   SELECT SUM(i.qty * i.price) AS revenue
  )
FROM order_revenue
SELECT ROUND(MIN(revenue)) AS minimum,
       ROUND(MAX(revenue)) AS maximum,
       ROUND(AVG(revenue)) AS average;
```

Result:

```
[ { "minimum": 130,
    "maximum": 18848,
    "average": 4670
  }
]
```

UNION ALL

In SQL, you can connect query blocks by using the set-oriented operators UNION, INTERSECT, and EXCEPT, each of which has ALL and DISTINCT versions, for a total of six operators. SQL has rules that define when two query blocks are similar enough to be combined in this way (it's called being *union-compatible*). But since N1QL operates on schemaless data, the result of a query block does not have a well-defined structure, and the query processor has no way to check query blocks for union compatibility. So in N1QL, any two query blocks can be combined by a set-oriented operator.

At present, N1QL supports only one of the set-oriented operators for combining query blocks: UNION ALL. The result of UNION ALL between two query blocks is a collection that includes all the items in the result of the first block and all the items in the result of the second block. Duplicate items are not eliminated from the query result.

In a query that is a UNION ALL of multiple query blocks, you get one more chance to define an ordering on the query result and to limit its size: you can include ORDER BY, LIMIT, or OFFSET clauses that apply to the query result as a whole. Of course, an ORDER BY clause will be most meaningful if the results generated by the query blocks have some field-names in common. An ORDER BY on the result of a UNION ALL expression overrules any ORDER BY clauses that may be found in individual query blocks inside the UNION ALL. In fact, an ORDER BY on an individual query block inside UNION ALL is meaningful only if the block also has a LIMIT and/or OFFSET clause; in that case, the ORDER BY, LIMIT, and OFFSET clauses work together to determine which and how many objects are returned by that query block.

The following example shows how information can be assembled by a union of query blocks on two separate datasets.

> (Q55) Make a list of the customer ids of customers who have an unknown zipcode or have placed an order for more than three different items.

```
SELECT c.custid, "Unknown zipcode" AS reason
FROM customers AS c
WHERE c.address.zipcode IS NOT KNOWN

UNION ALL
```

```
SELECT o.custid, "Big order" AS reason
FROM orders AS o
WHERE ARRAY_COUNT(o.items) > 3

ORDER BY reason;
```

Result:

```
[ { "custid": "C37",
    "reason": "Big order"
  },
  { "custid": "C47",
    "reason": "Unknown zipcode"
  }
]
```

You've now seen all the basic features of SQL++ as represented in N1QL for Analytics, and you're ready to strike out and write some queries of your own. Good luck, and have fun!

Appendix A:
Example Data

Appendix A: Example Data

The data in this Appendix was used to generate the results for the example queries in this tutorial.

All names, addresses, and other items in this example data are fictitious. Any resemblance between this data and actual persons, places, or events is coincidental and unintended.

Customers Dataset

```
[
  { "custid": "C13",
    "name": "T. Cruise",
    "address":
        { "street": "201 Main St.",
          "city": "St. Louis, MO",
          "zipcode": "63101"
        },
    "rating": 750
  },

  { "custid": "C25",
    "name": "M. Streep",
    "address":
        { "street": "690 River St.",
          "city": "Hanover, MA",
          "zipcode": "02340"
        },
    "rating": 690
  },

  { "custid": "C31",
    "name": "B. Pitt",
    "address":
        { "street": "360 Mountain Ave.",
          "city": "St. Louis, MO",
          "zipcode": "63101"
```

```
        }
    },

    { "custid": "C35",
      "name": "J. Roberts",
      "address":
          { "street": "420 Green St.",
            "city": "Boston, MA",
            "zipcode": "02115"
          },
      "rating": 565
    },

    { "custid": "C37",
      "name": "T. Hanks",
      "address":
          { "street": "120 Harbor Blvd.",
            "city": "Boston, MA",
            "zipcode": "02115"
          },
      "rating": 750
    },

    { "custid": "C41",
      "name": "R. Duvall",
      "address":
          { "street": "150 Market St.",
            "city": "St. Louis, MO",
            "zipcode": "63101"
          },
      "rating": 640
    },

    { "custid": "C47",
      "name": "S. Loren",
      "address":
          { "street": "Via del Corso",
            "city": "Rome, Italy"
```

```
          },
        "rating": 625
      }

   ]
```

Orders Dataset

```
   [
     { "orderno": 1001,
       "custid": "C41",
       "order_date": "2017-04-29",
       "ship_date": "2017-05-03",
       "items": [ { "itemno": 347,
                    "qty": 5,
                    "price": 19.99
                  },
                  { "itemno": 193,
                    "qty": 2,
                    "price": 28.89
                  }
                ]
     },

     { "orderno": 1002,
       "custid": "C13",
       "order_date": "2017-05-01",
       "ship_date": "2017-05-03",
       "items": [ { "itemno": 460,
                    "qty": 95,
                    "price": 100.99
                  },
                  { "itemno": 680,
                    "qty": 150,
                    "price": 8.75
                  }
                ]
     },
```

```
{ "orderno": 1003,
  "custid": "C31",
  "order_date": "2017-06-15",
  "ship_date": "2017-06-16",
  "items": [ { "itemno": 120,
              "qty": 2,
              "price": 88.99
            },
            { "itemno": 460,
              "qty": 3,
              "price": 99.99
            }
          ]
},

{ "orderno": 1004,
  "custid": "C35",
  "order_date": "2017-07-10",
  "ship_date": "2017-07-15",
  "items": [ { "itemno": 680,
              "qty": 6,
              "price": 9.99
            },
            { "itemno": 195,
              "qty": 4,
              "price": 35.00
            }
          ]
},

{ "orderno": 1005,
  "custid": "C37",
  "order_date": "2017-08-30",
  "items": [ { "itemno": 460,
              "qty": 2,
              "price": 99.98
            },
```

```
        { "itemno": 347,
          "qty": 120,
          "price": 22.00
        },
        { "itemno": 780,
          "qty": 1,
          "price": 1500.00
        },
        { "itemno": 375,
          "qty": 2,
          "price": 149.98
        }
      ]
  },

  { "orderno": 1006,
    "custid": "C41",
    "order_date": "2017-09-02",
    "ship_date": "2017-09-04",
    "items": [ { "itemno": 680,
                 "qty": 51,
                 "price": 25.98
               },
               { "itemno": 120,
                 "qty": 65,
                 "price": 85.00
               },
               { "itemno": 460,
                 "qty": 120,
                 "price": 99.98
               }
             ]
  },

  { "orderno": 1007,
    "custid": "C13",
```

```
        "order_date": "2017-09-13",
        "ship_date": "2017-09-20",
        "items": [ { "itemno": 185,
                     "qty": 5,
                     "price": 21.99
                   },
                   { "itemno": 680,
                     "qty": 1,
                     "price": 20.50
                   }
                 ]
      },

    { "orderno": 1008,
      "custid": "C13",
      "order_date": "2017-10-13",
      "items": [ { "itemno": 460,
                   "qty": 20,
                   "price": 99.99
                 }
               ]
    }

  ]
```

Appendix B:
Binding Variables
and Resolving Names

Appendix B: Binding Variables and Resolving Names

The most basic parts of N1QL expressions are mostly names. Names can appear in every clause of a query. Sometimes a name consists of a single identifier, such as `region` or `revenue`. Often a name consists of two identifiers separated by a dot, such as `customer.address`. Occasionally a name has more than two identifiers, such as `policy.owner.address.zipcode`. *Resolving* a name means finding exactly what the name refers to. This is usually pretty clear, but it's necessary to have well-defined rules for how to resolve a name in cases of ambiguity.

As we've seen, the basic job of each clause in a query block is to bind variables. Each clause sees the variables bound by previous clauses, and may bind some additional variables. Names are always resolved with respect to the variables that are bound ("in scope") at the place where the name occurs. It's always possible that the name resolution process will fail, which may lead to an `empty` result or an error message.

In this Appendix, we'll take a close look at how variables are bound and how names are resolved. If you just want to run some queries, you don't need to read this Appendix. Just define iteration variables in your FROM clauses and use the variables to qualify your field-names, and you will be fine. But if you are the type of person who really wants to know what is going on under the hood, hang on tight and read the rules.

One note before you start: When the system is reading a query and resolving names, it has a list of all the available dataverses and datasets. So it knows whether `a.b` is a valid name for dataset `b` in dataverse `a`. However, it does not have any knowledge about the schema of the data inside the datasets. So it does not know whether any object in a particular dataset has a field named `c`. These assumptions affect how errors are handled. If you try to access dataset `a.b` and no dataset by that name exists, you will get an error and your query will not run. But if you try to access a field `c` in a collection of objects, your query will run and return `missing` for each object that doesn't have a field named `c`.

Binding Variables

Variables can be bound in the following ways:

1. WITH and LET clauses bind a variable to the result of an expression in a straightforward way. Examples:

```
WITH cheap_parts AS (SELECT partno FROM parts WHERE price < 100)
```
binds the variable `cheap_parts` to the result of the subquery.

```
LET pay = salary + bonus
```
binds the variable `pay` to the result of evaluating the expression
`salary + bonus`.

2. FROM, GROUP BY, and SELECT clauses have optional AS subclauses that contain an
    expression and a name (called an *iteration variable* in a FROM clause, or an *alias* in
    GROUP BY or SELECT). Examples:

    ```
    FROM customer AS c, order AS o

    GROUP BY salary + bonus AS total_pay

    SELECT MAX(price) AS highest_price
    ```

An AS subclause always binds the name (as a variable) to the result of the expression (or,
in the case of a FROM clause, to the individual members of the collection identified by
the expression).

It's always a good practice to use the keyword AS when defining an alias or iteration
variable. However, the syntax allows the keyword AS to be omitted. For example, the
FROM clause above could have been written like this:

```
FROM customer c, order o
```

Omitting the keyword AS does not affect the binding of variables. The FROM clause in
this example binds variables c and o, whether the keyword AS is used or not.

In certain cases, a variable is automatically bound even if no alias or variable-name
is specified. Whenever an expression *could have been* followed by an AS subclause, if
the expression consists of a simple name or a path expression, that expression binds
a variable whose name is the same as the simple name or the last step in the path
expression. Here are some examples:

```
FROM customer, order
```
binds iteration variables named `customer` and `order`

```
GROUP BY address.zipcode
```
binds a variable named `zipcode`

```
SELECT item[0].price
```
binds a variable named `price`

Note that a FROM clause iterates over a collection (usually a dataset) and binds a variable to each member of the collection, in turn. The name of the collection remains in scope, but it is not a variable. For example, consider this FROM clause used in a self-join:

```
FROM customer AS c1, customer AS c2
```

This FROM clause joins the `customer` dataset to itself, binding the iteration variables c1 and c2 to objects in the left-hand side and right-hand side of the join, respectively. After the FROM clause, c1 and c2 are in scope as variables, and `customer` remains accessible as a dataset name but not as a variable.

3. Special rules for GROUP BY:

   a. If a GROUP BY clause specifies an expression that has no explicit alias, it binds a pseudo-variable that is lexicographically identical to the expression itself. For example:

   ```
   GROUP BY salary + bonus
   ```
   binds a pseudo-variable named `salary + bonus`.

   This rule allows subsequent clauses to refer to the grouping expression (`salary + bonus`) even though its constituent variables (`salary` and `bonus`) are no longer in scope. For example, the following query is valid:

   ```
   FROM employee
   GROUP BY salary + bonus
   HAVING salary + bonus > 1000
   ```

```
SELECT salary + bonus, COUNT(*) AS how_many
```

Although it might have been more elegant to simply require an explicit alias in cases like this, N1QL retains the pseudo-variable rule for SQL compatibility. Note that the expression `salary + bonus` is not actually evaluated in the HAVING and SELECT clauses (it could not be because `salary` and `bonus` are no longer in scope). Instead, `salary + bonus` is treated as a reference to the pseudo-variable defined in the GROUP BY clause.

b. A GROUP BY clause may be followed by a GROUP AS clause that binds a variable to the group. The purpose of this variable is to make the individual objects inside the group visible to subqueries that may need to iterate over them.

The GROUP AS variable is bound to a multiset of objects. Each object represents one of the members of the group. Since the group might have been formed from a join, each of the member-objects contains a nested object for each variable bound by the nearest FROM clause (and its LET subclause, if any). These nested objects, in turn, contain the actual fields of the group-member. To understand this process, consider the following example:

```
FROM parts AS p, suppliers AS s
WHERE p.suppno = s.suppno
GROUP BY p.color GROUP AS g;
```

Suppose that the objects in parts have fields `partno, color,` and `suppno`. Suppose that the objects in `suppliers` have fields `suppno` and `location`.

Then, for each group formed by the GROUP BY, the variable g will be bound to a multiset with the following structure:

```
[ { "p": { "partno": "p1", "color": "red", "suppno": "s1" },
    "s": { "suppno": "s1", "location": "Denver" } },
  { "p": { "partno": "p2", "color": "red", "suppno": "s2" },
```

```
            "s": { "suppno": "s2", "location": "Atlanta" } },
        . . .
    ]
```

Scoping

In general, the variables that are in scope at a particular position are variables that were bound earlier in the current query block, in outer (enclosing) query blocks, or in the WITH clause at the beginning of the query. More specific rules follow.

The clauses in a query block are conceptually processed in the following order:

FROM (followed by LET subclause, if any)
WHERE
GROUP BY (followed by LET subclause, if any)
HAVING
SELECT or SELECT VALUE
ORDER BY
OFFSET
LIMIT

During processing of each clause, the variables that are in scope are those variables that are bound in the following places:

1. In earlier clauses of the same query block (as defined by the ordering given above).
    Example: `FROM orders AS o SELECT o.date`
    The variable o in the SELECT clause is bound, in turn, to each object in the dataset `orders`.

2. In outer query blocks in which the current query block is nested. In case of duplication, the innermost binding wins.

3. In the WITH clause (if any) at the beginning of the query.

However, in a query block where a GROUP BY clause is present:

1. In clauses processed before GROUP BY, scoping rules are the same as though no GROUP BY were present.

2. In clauses processed after GROUP BY, the variables bound in the nearest FROM-clause (and its LET subclause, if any) are removed from scope and replaced by the variables bound in the GROUP BY clause (and its LET subclause, if any). However, this replacement does not apply inside the arguments of the five SQL special aggregating functions (MIN, MAX, AVG, SUM, and COUNT). These functions still need to see the individual data items over which they are computing an aggregation. For example, after `FROM employee AS e GROUP BY deptno`, it would not be valid to reference `e.salary`, but `AVG(e.salary)` would be okay.

Special case: In an expression inside a FROM clause, a variable is in scope if it was bound in an earlier expression in the same FROM clause. Example:

```
FROM orders AS o, o.items AS i
```

The reason for this special case is to support iteration over nested collections.

Note that since the SELECT clause comes after the WHERE and GROUP BY clauses in conceptual processing order, any variables defined in SELECT are not visible in WHERE or GROUP BY. Therefore, the following query will not return the expected result (in the WHERE clause, `pay` will be interpreted as a field in the `emp` object rather than as the computed value `salary + bonus`):

```
SELECT name, salary + bonus AS pay
FROM emp
WHERE pay > 1000
ORDER BY pay
```

The probable intent of the query above can be accomplished as follows:

```
FROM emp AS e
LET pay = e.salary + e.bonus
WHERE pay > 1000
SELECT e.name, pay
ORDER BY pay
```

Resolving Names

The process of name resolution begins with the leftmost identifier in the name. The rules for resolving the leftmost identifier are as follows:

1. <u>In a FROM clause:</u> Names in a FROM clause identify the collections over which the query block will iterate. These collections may be stored datasets, or may be the results of nested query blocks. A stored dataset may be in a named dataverse or in the default dataverse. Thus, if the two-part name `a.b` is in a FROM clause, a might represent a dataverse and b might represent a dataset in that dataverse. Another example of a two-part name in a FROM clause is `FROM orders AS o, o.items AS i`. In `o.items`, `o` represents an `order` object bound earlier in the FROM clause, and `items` represents the `items` object inside that `order`.

    The rules for resolving the leftmost identifier in a FROM clause (including a JOIN subclause), or in the expression following IN in a quantified predicate, are as follows:

    a. If the identifier matches a variable-name that is in scope, it resolves to the binding of that variable. (Note that in the case of a subquery, an in-scope variable might have been bound in an outer query block; this is called a *correlated subquery*.)

    b. Otherwise, if the identifier is the first part of a two-part name like `a.b`, the name is treated as dataverse.dataset. If the identifier stands alone as a one-part name, it is treated as the name of a dataset in the default dataverse. An error will result if the designated dataverse or dataset does not exist.

2. <u>Elsewhere in a query block:</u> In clauses other than FROM, a name typically identifies a field of some object. For example, if the expression `a.b` is in a SELECT or WHERE clause, it's likely that a represents an object and b represents a field in that object.

The rules for resolving the leftmost identifier in clauses other than the ones listed in Rule 1 are as follows:

a. If the identifier matches a variable-name that is in scope, it resolves to the binding of that variable. (In the case of a correlated subquery, the in-scope variable might have been bound in an outer query block.)

b. (The "Single Variable Rule"): Otherwise, if the FROM clause (and the LET clause that follows it, if any) in the current query block binds exactly one variable, the identifier is treated as a field access on the object bound to that variable. For example, in the query `FROM customer SELECT address`, the identifier `address` is treated as a field in the object bound to the variable `customer`. At runtime, if the object bound to `customer` has no `address` field, the `address` expression will return `missing`. If the FROM clause (and its LET subclause, if any) in the current query block binds multiple variables, name resolution fails with an "ambiguous name" error. Note that the Single Variable Rule searches for bound variables only in the current query block, not in outer (containing) blocks. The purpose of this rule is to permit N1QL to resolve field-references unambiguously without relying on any schema information.

Exception: In a query that has a GROUP BY clause, the Single Variable Rule does not apply in any clauses that occur after the GROUP BY because, in these clauses, the variables bound by the FROM clause are no longer in scope. In clauses after GROUP BY, only Rule 2a applies.

3. In an ORDER BY clause following a UNION ALL expression:

The leftmost identifier is treated as a field-access on the objects that are generated by the UNION ALL. For example:

```
query-block-1
UNION ALL
query-block-2
ORDER BY zipcode;
```

In the result of this query, objects that have a `zipcode` field will be ordered by the value of this field; objects that have no `zipcode` field will appear at the beginning of the query result (in ascending order) or at the end (in descending order).

4. Once the leftmost identifier has been resolved, the following dots and identifiers in the name (if any) are treated as a path expression that navigates to a field nested inside that object. The name resolves to the field at the end of the path. If this field does not exist, the value `missing` is returned.

Appendix C: Running the Example Queries

Appendix C: Running the Example Queries

This Appendix will explain how to load the example data from Appendix A into Couchbase and run the example queries using the Analytics Service. The Analytics Service is part of Couchbase Enterprise Edition, which you can download freely from this URL:

https://www.couchbase.com/downloads

The materials you need to load the data and run the queries can be found on the Tutorial Materials page at this URL:

https://www.couchbase.com/n1ql-for-analytics

First, some background. Couchbase Server provides multiple services, two of which are the Query Service and the Analytics Service. The Query Service provides a direct way to query and update your operational data using N1QL for Query. Operational data is automatically replicated on the Analytics Service, where it can be queried (but not updated) using N1QL for Analytics. The two versions of N1QL are similar but not identical. N1QL for Analytics is the Couchbase implementation of SQL++.

In order to use N1QL for Analytics, the Analytics Service must be enabled in your Couchbase installation. If you are installing Couchbase for the first time, make sure that you configure your system to enable the Analytics Service (this may not be the default on your platform.)

To run the examples in this tutorial, you must first load the example data into your operational database. The best way to do this is by using the Query Service, following these steps:

1. Start the Couchbase Web Console and look at the Dashboard display. On the left side of the display, you will see a list of services that can be invoked from the Dashboard, including Buckets, Query, and Analytics.

2. Click on Buckets and use the Add Buckets feature to create two new buckets called `customers` and `orders`. For each bucket, set Bucket Type to Couchbase and set the Memory Quota to the minimum (100MB).

3. Returning to the Dashboard, click on Query to invoke the Query Service. On the Tutorial Materials page (URL above), find the two SQL INSERT statements that begin with "`insert into customers`" and "`insert into orders`". Copy and paste these statements, one at a time, into the Query Editor and click Execute. The sample data is now loaded into your operational database.

4. If you wish to query the sample data on the Query Service, each bucket must have a primary index. You can accomplish this by executing the following statements on the Query Service:

```
create primary index on customers;
create primary index on orders;
```

These indexes are used only by the Query Service. If you plan to query your data by the Analytics Service only, you can skip this step.

The example queries in the tutorial are written in N1QL for Analytics and are intended to run on the Analytics Service. Some of them will not run on the Query Service due to syntactic differences between N1QL for Analytics and N1QL for Query. If you wish to query the example data on the Query Service, you can write your own queries using N1QL for Query.

5. Returning to the Dashboard, click on Analytics to invoke the Analytics Service. Type the following statements into the Analytics Query Editor and execute them:

```
create dataset on customers;
create dataset on orders;
connect link Local;
```

These statements will cause your operational data to be replicated from the `customers` and `orders` buckets onto the Analytics Service, where you can query it using N1QL for Analytics.

6. Returning to the Tutorial Materials page, you can now copy and paste each of the example queries into the Analytics Query Editor and click Execute to run them. You can also modify the example queries or make up some new queries of your own.

7. If you wish to remove the example data from your system, just retrace your steps, as follows:

On the Analytics Service:

```
disconnect link Local;
drop dataset customers;
drop dataset orders;
```

On the Dashboard, click Buckets, then click on each of the example buckets (customers and orders) and click Delete for each bucket.

Made in the USA
Las Vegas, NV
21 July 2021